Biosystems & Biorobotics

Volume 24

Eugenio Guglielmelli, Laboratory of Biomedical Robotics,
Campus Bio-Medico University of Rome,
Rome, Romania   *Series Editor*

The BIOSYSTEMS & BIOROBOTICS (BioSysRob) series publishes the latest research developments in three main areas: 1) understanding biological systems from a bioengineering point of view, i.e. the study of biosystems by exploiting engineering methods and tools to unveil their functioning principles and unrivalled performance; 2) design and development of biologically inspired machines and systems to be used for different purposes and in a variety of application contexts. In particular, the series welcomes contributions on novel design approaches, methods and tools as well as case studies on specific bio-inspired systems; 3) design and developments of nano-, micro-, macro- devices and systems for biomedical applications, i.e. technologies that can improve modern healthcare and welfare by enabling novel solutions for prevention, diagnosis, surgery, prosthetics, rehabilitation and independent living. On one side, the series focuses on recent methods and technologies which allow multi-scale, multi-physics, high-resolution analysis and modeling of biological systems. A special emphasis on this side is given to the use of mechatronic and robotic systems as a tool for basic research in biology. On the other side, the series authoritatively reports on current theoretical and experimental challenges and developments related to the "biomechatronic" design of novel biorobotic machines. A special emphasis on this side is given to human-machine interaction and interfacing, and also to the ethical and social implications of this emerging research area, as key challenges for the acceptability and sustainability of biorobotics technology. The main target of the series are engineers interested in biology and medicine, and specifically bioengineers and bioroboticists. Volume published in the series comprise monographs, edited volumes, lecture notes, as well as selected conference proceedings and PhD theses. The series also publishes books purposely devoted to support education in bioengineering, biomedical engineering, biomechatronics and biorobotics at graduate and post-graduate levels.

Indexed by SCOPUS and Springerlink. The books of the series are submitted for indexing to Web of Science.

More information about this series at ▶ http://www.springer.com/series/10421

Michael Chappell
Stephen Payne

Physiology for Engineers

Applying Engineering Methods
to Physiological Systems

Second Edition

 Springer

Michael Chappell
Beacon of Excellence in Precision
Imaging & Sir Peter Mansfield
Imaging Centre, School of Medicine
University of Nottingham
Nottingham, UK

Stephen Payne
Department of Engineering Science
University of Oxford
Oxford, UK

ISSN 2195-3562   ISSN 2195-3570   (electronic)
Biosystems & Biorobotics
ISBN 978-3-030-39707-4   ISBN 978-3-030-39705-0   (eBook)
https://doi.org/10.1007/978-3-030-39705-0

This Springer imprint is published by the registered company Springer Nature Switzerland AG
The registered company address is: Gewerbestrasse 11, 6330 Cham, Switzerland

To Angela Chappell BSc (Hons) aka Mum

Michael Chappell

To Alan

Stephen Payne

Preface

As its name indicates, this short book is intended to give students in any branch of engineering an introduction to human physiology and how it can be approached by those with an engineering background. It will also provide an introduction for students in other physical science subjects to the approaches that can be adopted in understanding and modelling human physiology. We hope that it will provide a starting point for both students and non-students to explore what is fascinating about human **physiology** (which simply means the study of nature).

This book was written particularly with those wanting to enter the field of biomedical engineering in mind. As biomedical engineers ourselves, we often get asked what biomedical engineering actually is and what its purpose is. Neither of us set out to train as biomedical engineers (we did our final year undergraduate projects on land mines and jet engines), but we both ended up in this area because it opened up so many interesting problems to work on. If engineering is the "appliance of science"[1], then biomedical engineering is the appliance of engineering to medicine. The whole point is to help to solve problems that will one day improve healthcare.

Let's give you an example, directly related to our research.

How can we best decide what treatment to give someone coming into a hospital with a stroke? Doctors will do brain imaging, mostly likely a CT scan, or in some hospitals an MRI scan (we'll explain what these are later in the book). How does the doctor then decide what best to do? Stroke is essentially a plumbing problem and comes in two forms:
Blockage—an ischaemic stroke arises when a vessel feeding part of the brain gets blocked.
Leakage—an haemorrhagic stroke is when blood is leaking into the brain from a broke blood vessel.
Both forms of stroke need urgent attention and the solutions seem simple: remove a blockage or stop a leak.

However, there are lots of complications to this. For example, in an ischaemic stroke, should the doctor operate and try to pull the clot out or should he/she try to dissolve the clot by injecting a blood thinning drug? Often the person is very elderly with many other illnesses and so an operation is not a good idea; likewise the blood thinner may result in a blood vessel bursting and a haemorrhage (leak) elsewhere in the brain.

---

1 Our thanks to Zanussi.

The doctor has to make a rapid treatment decision and does so based on a lot of experience. What we can do as biomedical engineers in this context is to help in the decision making process by providing (1) the best and most useful information from the brain imaging; and (2) the best predictions of outcome for various treatment decisions. By doing this and providing the doctor with the most useful information to make a well-informed decision, biomedical engineers can help to get the best outcomes for these patients. This will not only help to save lives, but also help to reduce the levels of disability that stroke patients have to live with when they leave the hospital. Even a small improvement in the outcome of a stroke patient can make the difference between living at home and living in a home, or between independent and dependent living.

The tools we have to do this are both physical (imaging devices) and mathematical (systems of equations), things that engineers excel in. These need to be matched with an understanding of the problem and this requires a knowledge not only of how the body works, but also how our tools apply to that physiology. Hence this book: an attempt to guide the reader in bridging what appears to be a divide between physical and biomedical sciences.

There are a number of things to note about the book:
1. It is **introductory**: deliberately we are covering a wide range of material at a high level. There are many other very good text books that will take you into much greater detail, but the purpose here is simply to give you an introduction to physiology from an engineering viewpoint. We also do not attempt to cover all of physiology, rather we have chosen what we hope are a selection of illustrative and interesting examples.
2. It is **quantitative**: the focus is very much on turning the underlying physiology into mathematics so that a system, whether it is a single cell or the whole brain, can be modelled and interpreted using engineering techniques.
3. It is concerned with clinical **measurements**: since we can only know about what we can measure, throughout the book we have detailed the most common clinical measurement techniques that can be used to gain information about the topics presented here.

At the end of the book, we hope that you will have gained some insight into human physiology and how we can express this mathematically and measure it, using core engineering techniques. You will then hopefully also be better equipped to apply the same principles and techniques to other aspects of physiology that we haven't covered here.

The book starts with the cell, which is the fundamental unit of the body. We will examine its structure, its function and how it operates: in particular we will examine the generation of the action potential and how this is transmitted between cells. We will then gradually move to larger scales and look at various systems in the body, primarily the cardiovascular, respiratory and nervous systems.

We have included exercises, but instead of including them all at the end of each chapter, we have deliberately inserted them in the text at the most appropriate point. Many of the examples require you to explore a result that has been stated or to work through a similar example. It is not necessary to do the examples as you read through the book (and they are clearly highlighted), but we hope that they will be helpful in reinforcing the ideas presented, and in particular giving you experience in applying engineering techniques to physiological problems. There are brief solutions given at the end of the book.

This book does draw on a very wide range of engineering knowledge, which we appreciate that you will not all have. For some of you phasor analysis will be a mystery, for others elasticity theory will be incomprehensible. We have tried to keep the level of knowledge required to a minimum, but since this is not a course in basic engineering, there may be some parts that you will not understand. We hope that at least you will get some insight into what is going on, even if you don't follow all of the equations and mathematics in every chapter.

As academics, we have taught a similar course for nearly ten years to third and fourth year MEng undergraduate and MSc graduate students at the University of Oxford. We have, like all academics, learnt a great deal ourselves over this time and we take this opportunity to thank the students who have attended our course in its various different incarnations. In all of this we have been able to draw on our experience teaching our students, to the book's great benefit; however, any errors that remain are naturally entirely our fault.

Michael Chappell
Stephen Payne
Oxford, UK
March 2020

Contents

Cell Structure and Biochemical Reactions

Contents

© Springer Nature Switzerland AG 2020
M. Chappell and S. Payne, *Physiology for Engineers*, Biosystems & Biorobotics 24,
https://doi.org/10.1007/978-3-030-39705-0_1

1

In this chapter, we will start by examining the cell, since it is the fundamental unit of living matter. We will consider what it is made up of as we will need to consider this when we look at how it operates. We will then also consider how to model biochemical reactions as we will also need this to model how the cell operates. This will then form the basis for the remainder of the chapters in this book.

1.1 Cell Structure

There are brain cells, heart cells, liver cells and so on, in every part of the body. Each of these cells has a specific function and purpose. Despite this, all cells have some characteristics in common, as shown schematically in ◘ Fig. 1.1. There are four main components. Cells have an outer layer, called the **membrane**, which acts as the boundary between the inside and outside of the cell. We will examine this in more detail in the next chapter.

Inside the cell, there is a **nucleus** that contains the cell's DNA, i.e. the genetic code that determines what the cell does, and many small structures that carry out the operations for the cell, termed **organelles**. Unsurprisingly there are lots of these, each with

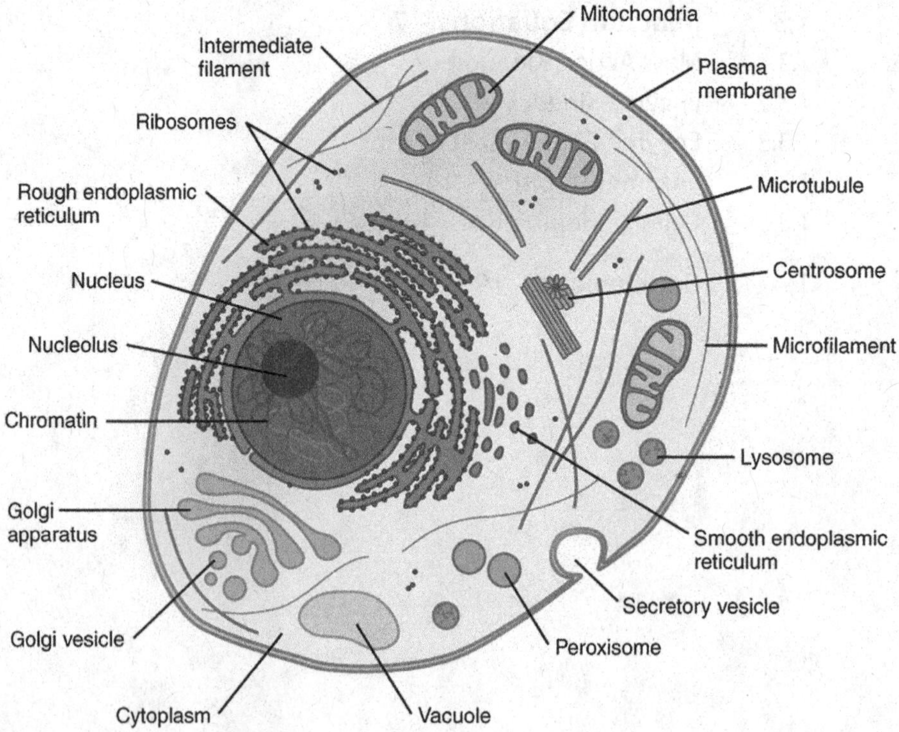

◘ Fig. 1.1 Structure of the cell (this figure is taken, without changes, from OpenStax College under license: ► http://creativecommons.org/licenses/by/3.0/)

particular roles. They include ribosomes, lysosomes and mitochondria, where many of the reactions that produce energy take place. The rest of the cell is occupied by a gel-like fluid, called the **cytosol**, that contains ions and other substances and that surrounds the other internal elements. The cytosol and the organelles together are termed the **cytoplasm**. We will examine a number of these elements, primarily the membrane and the cytoplasm, in more detail later on.

More formally, we classify a cell as the **smallest independently viable unit** of a living organism. It is therefore self-contained and self-maintaining. The smallest organism is made up of just one cell, whereas the largest are made up of billions of cells. All cells can **reproduce** by cell division and **metabolise**, which means that they take in and process raw materials and release the by-products of metabolism. Cells also respond to both external and internal stimuli, for example temperature and pH.

1.2 Cell Chemicals

Before examining the cell in detail, we will look at the different types of chemicals found in the body and their roles, such that when we consider the functions of the cell, it will be easier to understand what it is doing. Note that we will only look at a few of the most important substances found within a cell.

The important chemicals within the body can be divided into two categories: **inorganic** and **organic**. The difference is very simply that organic compounds always contain carbon, whereas most inorganic compounds do not contain carbon.

The main **inorganic** substances are:
- Water, which acts as a solvent, a biochemical reactant, a regulator of body temperature and a lubricant;
- Electrolytes, which balance osmotic pressure and biochemical reactants;
- Acids and bases, which act to balance pH.

The primary **organic** substances are:
- Carbohydrates;
- Lipids;
- Proteins;
- Nucleic acids (including DNA and RNA);
- Adenosine triphosphate (ATP).

Carbohydrates, proteins and nucleic acids are all necessary for life to function.

Cells are made up of just three substances: water; inorganic ions; and organic molecules. Water makes up around 70% of cell mass, with most of the rest being organic molecules: inorganic ions make up only a few percent. The inorganic ions include sodium (Na^+), potassium (K^+), calcium (Ca^{2+}), chloride (Cl^-) and bicarbonate (HCO_3^-). The plus and minus signs are used to tell us that the ions are positively or negatively charged: this will be very important when we consider how the cell behaves later on. We don't always explicitly write the plus or minus signs though, since only a handful of ions are involved and we tend to know what their charges are.

1

1.2.1 Proteins

There are many types of **proteins** and they perform lots of roles within the body, including:

- Structural (coverings and support);
- Regulatory (hormones, control of metabolism);
- Contractile (muscles);
- Immunological (antibodies, immune system);
- Transport (movement of materials, haemoglobin for oxygen);
- Catalytic (enzymes).

Proteins are large molecules made up of amino acids (of which there are around 20 in the human body): examples of proteins are **insulin**, **haemoglobin** and **myosin**. Insulin plays a key role in the regulation of blood sugar levels, haemoglobin is vital to the transport of oxygen around the body, and myosin is used in muscle contraction.

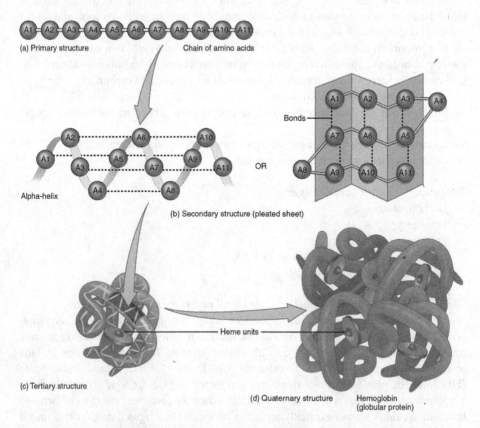

☐ **Fig. 1.2** Protein structure of haemoglobin (this figure is taken, without changes, from OpenStax College under license: ▶ http://creativecommons.org/licenses/by/3.0/)

Proteins differ between themselves in their order of amino acids, which is determined by their gene sequence; they are folded into specific three-dimensional shapes. The unique folded structure of a protein is often a very important determinant of its function. Proteins are formed, used and recycled with a lifespan that can vary from minutes to years. A schematic of the structure of haemoglobin, a typical protein, is shown in ◘ Fig. 1.2. The final structure is made up of four sub-units, each of which is a highly folded and bonded version of the original chain of amino acids.

1.2.2 ATP

Adenosine triphosphate (**ATP**) is essentially the energy source for cells and acts like a battery. It is created by the conversion of glucose and oxygen into carbon dioxide, water and ATP:

$$C_6H_{12}O_6 + 6O_2 \rightarrow 6CO_2 + 6H_2O + ATP \qquad (1.1)$$

This is why we breathe in oxygen and breathe out carbon dioxide: to convert glucose (a sugar) into an energy store that can be used for cells to function.

◘ Figure 1.3 shows the three parts that make ATP molecules, with the third part comprising three phosphate groups (hence 'triphosphate'). When the third, terminal, phosphate is released, energy is liberated: ATP then becomes **ADP** (adenosine diphos-

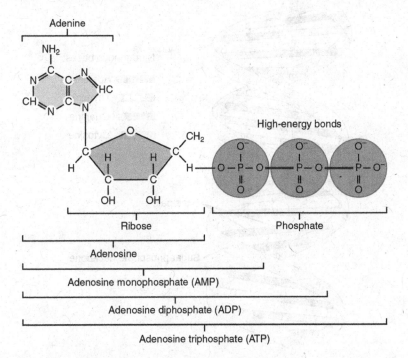

◘ **Fig. 1.3** Structure of ATP (this figure is taken, without changes, from OpenStax College under license: ► http://creativecommons.org/licenses/by/3.0/)

1

phate), i.e. with only two phosphate groups. If the second phosphate is also released, to yield **AMP** (adenosine monophosphate), more energy is liberated. This energy is used to power many cellular processes, as we will see later.

1.2.3 DNA/RNA

Deoxyribonucleic acid (DNA for short) is the molecule that contains most of the genetic information that controls all of the processes that occur in the cell. As can be seen in ◘ Fig. 1.4, there are two base pairs: adenine (A) with thymine (T), and guanine (G) with cytosine (C). These are connected to a sugar and a phosphate molecule, together called a nucleotide, which are arranged in the famous double helix structure shown in ◘ Fig. 1.4. The order of the bases encodes the information contained in each strand of DNA.

DNA is then organised into **chromosomes**: human cells have 23 pairs of chromosomes including the X and Y sex chromosomes. Chromosomes are sub-divided into genes, with each gene having its own function, instructing cells to make specific proteins. Ribonucleic acid (RNA) is the molecule that is used to convey genetic information by translating the information in a specific gene into a protein's amino acid sequence: unlike DNA, RNA is single-stranded.

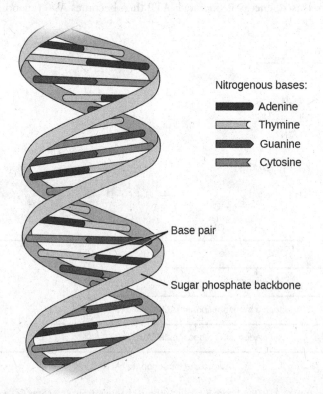

Nitrogenous bases:

■▶ Adenine
I⊏ Thymine
■▶ Guanine
■◁ Cytosine

Base pair

Sugar phosphate backbone

◘ **Fig. 1.4**   Structure of DNA (this figure is taken, without changes, from OpenStax College under license: ▶ http://creativecommons.org/licenses/by/3.0/)

1.3 Reaction Equations

Now that we have discussed how a cell is constructed, it is time to consider how we analyse its behaviour. We do this analysis using reaction equations, which describe how one set of atoms and molecules (**reactants**) combine to form a different set of atoms and molecules (**products**).

However, before we start to do this, we need to consider the correct use of units. Throughout this book, we will use the **mole**, which is formally defined as the number of atoms found in 12 g of ^{12}C (6.022×10^{23} atoms), as the basic unit. Chemical equations are all based on the use of the mole, since it is a much more convenient means of describing quantities than simple weight or volume.

We said earlier that 70% of the cell is water and so for many physiological elements, the molecules are not found in isolation, but in solution, primarily in water. We thus need a way to describe how much of the **solute** is present in the solution. The most common definition, although there are others, is called **molarity**: this is the number of moles of the solute per litre of solution. It thus has units of mol/l, which is normally written in shorthand as M. We will use this all the way through the remainder of this book unless stated otherwise. Since most physiological molarities are much less than 1 M, you should get used to seeing the abbreviation for milli-molar, which is written mM.

1.3.1 Mass Action Kinetics

Let's start by writing down one of the very simplest reaction equations:

$$A + B \xrightarrow{k} C \tag{1.2}$$

k is termed the **rate constant** for this reaction, which simply takes two reactants, A and B, and converts them into one product C. The quantity of C increases dependent upon the quantities of both A and B, thus a simple model for rate of change of the concentration of the product with time (termed the **reaction rate**) is given by:

$$\frac{d[C]}{dt} = k[A][B] \tag{1.3}$$

This is called the law of mass action and systems that obey this style of equation are said to be governed by **mass action kinetics**. The rate constant will depend upon the sizes and shapes of A and B; it also varies strongly with temperature (as well as with other factors such as pH).

Mass action kinetics are based on C being produced when A and B collide and combine, so-called **elementary reactions**. The rate constant is thus proportional to the number of collisions between A and B per unit time and the probability that the collision has enough energy.

Note that concentrations are usually denoted by square brackets, i.e. the concentration of A is $[A]$. However, for simplicity we will often use lower case letters to refer to chemical concentrations where we need to write many equations (see the next example below, where we do this to make writing the equations easier).

We should also note that Eq. (1.3) is not always true: for very high or very low concentrations, the rate of change is limited by other factors. Despite this, the equation is a good first approximation and will enable us to analyse even quite complicated systems in a straightforward way.

If there are multiple moles on the left hand side of the reaction equation, then the reaction rate is proportional to the concentrations to the relevant powers, i.e. the reaction rate for the reaction:

$$A + 2B \xrightarrow{k} C \tag{1.4}$$

is:

$$\frac{dc}{dt} = kab^2 \tag{1.5}$$

Also, all reactions strictly speaking must be reversible to some extent, thus there are forward and reverse rate constants, which need not be the same:

$$A + B \underset{k_-}{\overset{k_+}{\rightleftharpoons}} C \tag{1.6}$$

For this example, we can write down three equations, one for each of the chemical substances:

$$\frac{da}{dt} = k_-c - k_+ab \tag{1.7}$$

$$\frac{db}{dt} = k_-c - k_+ab \tag{1.8}$$

$$\frac{dc}{dt} = k_+ab - k_-c \tag{1.9}$$

Although this now looks rather complicated, if we add the first and third equations together, the rate of change of $a+c$ is equal to zero. Integrating up this equation tells us that the sum of these two concentrations is a constant, which we define here as: $a+c=a_o$, where a_o can be thought of as the initial amount of A before any was converted to C.

We can use the **equilibrium** condition in these equations. By setting the rates of change to zero in the three equations, we find:

$$\frac{k_-}{k_+} = \frac{\bar{a}\bar{b}}{\bar{c}} = K \tag{1.10}$$

where we use the over-bar to refer to the equilibrium or steady-state values of each substance. The ratio of the two rate constants, which we call K, is termed the **equilibrium constant**. This measures the relative preference for the substances to be in the combined, C, or separate, $A+B$, state. Note that we can define this either way up: we have chosen here to use the backwards rate divided by the forward rate, but we could just as easily have defined it the other way.

This equation can be used to determine the equilibrium constant using the equilibrium values of the concentrations. The rule for determining the constant is to multiply all the reactant concentrations to the power of their coefficients and divide by the product of the product concentrations, again each to the power of their coefficients. The following two exercises will help you get some practice at this.

Exercise A

A reaction takes place according to:

$$A + 2B \underset{k_-}{\overset{k_+}{\rightleftharpoons}} C + D \tag{A.1}$$

Write down the four reaction equations and show that the equilibrium constant is given by:

$$\frac{k_-}{k_+} = \frac{\bar{a}\bar{b}^2}{\bar{c}\bar{d}} \tag{A.2}$$

Exercise B

A reaction takes place that turns reactants A and B into products D, E and F according to the following separate reaction steps:

$$A + B \underset{k_{-1}}{\overset{k_{+1}}{\rightleftharpoons}} C + D \tag{B.1}$$

$$C \underset{k_{-2}}{\overset{k_{+2}}{\rightleftharpoons}} E + F \tag{B.2}$$

a. Write down the reaction equations for A using Eq. B.1 and for C using Eq. B.2.
b. Assuming that both reactions are in equilibrium, show that the equilibrium constant for the whole reaction is given by:

$$K = \frac{\bar{a}\bar{b}}{\bar{d}\bar{e}\bar{f}} = \frac{k_{-1}}{k_{+1}} \frac{k_{-2}}{k_{+2}} \tag{B.3}$$

i.e. the overall equilibrium constant is the product of the two individual equilibrium constants.

1.3.2 Enzyme Kinetics

Now that we have considered simple reaction equations and how to analyse them, we will look at one particular category of reaction that is very important: one where the reaction is catalysed by an **enzyme**. Enzymes are essentially substances that help other molecules, known as **substrates**, to change into products but which themselves are left unaffected by the reaction: they are sometimes known as **catalysts**.

1

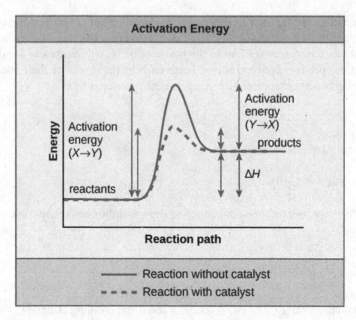

■ **Fig. 1.5**   Enzymes and activation energy (this figure is taken, without changes, from OpenStax College under license: ▶ http://creativecommons.org/licenses/by/3.0/)

One way in which enzymes work is by lowering the **activation energy** of the reaction, i.e. they make it easier to move from one state to another, as shown in ■ Fig. 1.5. Activation energy is the transient energy required to proceed with a reaction and is often larger than the energy required or released by the reaction itself. For example, a reaction may require breaking of some atomic bonds before new bonds are formed. Thus to proceed along the reaction, moving from left to right in the figure, a large and temporary source of energy is required, giving the characteristic 'hill' shown here. The enzyme lowers this 'hill', making it easier for the reaction to proceed.

Enzymes are particularly efficient at speeding up biological reactions and are highly specific, thus allowing very precise control of the reaction speed. Remember that our simple model is that of an elementary reaction that depends upon the rate of collision of the reactants and the probability of the reaction having enough energy. A simple example of enzyme action would be a protein that 'fits' a particular molecule in such a way that it causes a bond to be stressed making that bond more easily broken and thus reducing the activation energy, making the probability of reaction higher. Alternatively a protein might have 'sites' to which the species involved in the reaction bind, so that the enzyme increases the rate at which the species are brought together.

The first model to consider enzyme reactions was proposed by Michaelis and Menten. The enzyme E converts the substrate S into the product P in two stages. S and E first combine to give a complex C (sometimes written as ES), which then breaks down into E and P:

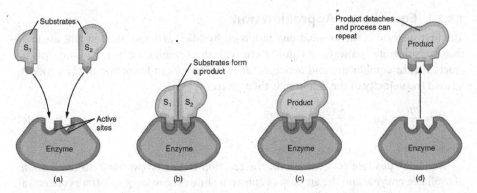

Fig. 1.6 Stages in an enzyme reaction (this figure is taken, without changes, from OpenStax College under license: ▶ http://creativecommons.org/licenses/by/3.0/)

$$E + S \underset{k_{-1}}{\overset{k_{+1}}{\rightleftharpoons}} C \overset{k_{+2}}{\rightarrow} E + P \tag{1.11}$$

In theory this second reaction can also work backwards, but normally P is continually removed, which prevents the reverse reaction from occurring, so we approximate it as a one way reaction. We can represent this process in diagrammatic form, as shown in ▶ Fig. 1.6, where in this case we have two substrates (S_1 and S_2) that both join together with the enzyme to form a single product before being released from the enzyme.

We can write down the four differential equations in the same way as before for the four different substances, S, E, C and P:

$$\frac{ds}{dt} = k_{-1}c - k_{+1}se \tag{1.12}$$

$$\frac{de}{dt} = k_{-1}c - k_{+1}se + k_{+2}c \tag{1.13}$$

$$\frac{dc}{dt} = k_{+1}se - k_{+2}c - k_{-1}c \tag{1.14}$$

$$\frac{dp}{dt} = k_{+2}c \tag{1.15}$$

Note that this set of equations is redundant: the second and third equations add to give zero, which means that the sum of E and C is constant over time and so we introduce a new constant, normally represented by $e + c = e_o$.

There are two common approaches to analysing this system of equations: the equilibrium approximation and the quasi-steady-state approximation. We will examine them both briefly here to give an insight into how we analyse more complex sets of reaction equations.

1

1.3.2.1 Equilibrium Approximation

The first method is the original one proposed by Michaelis and Menten and assumes that the substrate is always in equilibrium with the complex, i.e. the first stage of the reaction is in equilibrium and hence $ds/dt = 0$. The rate of formation of the product, termed the **velocity** of the reaction, is then given by:

$$V = \frac{dp}{dt} = k_{+2}c = \frac{k_{+2}e_o s}{K_s + s} \tag{1.16}$$

where $K_s = k_{-1}/k_{+1}$, similarly to before.

At small substrate concentrations, the reaction rate is proportional to the amount of available enzyme and the amount of substrate: however, at large substrate concentrations, the rate is limited by the amount of enzyme present. The second reaction is thus termed **rate limiting** since the reaction velocity cannot increase beyond a certain value.

Equation (1.16) is often re-written in the form:

$$V = \frac{V_{max}s}{K_s + s} \tag{1.17}$$

where V_{max} is so called because it is the maximum velocity with which the reaction can proceed. This equation is known as the **Michaelis-Menten equation** and is used widely in physiological modelling to mimic processes where there is a limited rate of reaction. It is shown in ☐ Fig. 1.7, where the asymptote and the initial slope are shown as dotted lines: note that the square marks the point at which the reaction velocity reaches half of its maximum value (this is obviously when $s = K_s$).

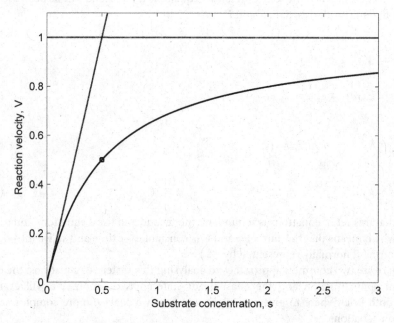

☐ **Fig. 1.7**   Michaelis-Menten reaction behaviour (for $K_s = 0.5$ and $V_{max} = 1$)

1.3.2.2 Quasi-Steady-State Approximation

The second approximation assumes that the rates of formation and the breakdown of the complex are equal, thus $dc/dt = 0$. Solution of the remaining equations gives a very similar result to the previous reaction velocity:

$$V = \frac{V_{max}s}{K_m + s} \tag{1.18}$$

where $K_m = (k_{-1} + k_2)/k_{+1}$. Clearly the two approximations give very similar results, and they are both of Michaelis-Menten form, but they are based on very different assumptions.

Exercise C
Derive the results show in Eqs. (1.17) and (1.18). Explain why the quasi-steady-state reaction velocity is always smaller than the equilibrium reaction velocity.

Exercise D
Explain why plotting $1/V$ as a function of $1/s$ for a Michaelis-Menten reaction gives a straight line. Sketch this function and explain how the constants in this equation can be found from the intercepts with the two axes.

This model is very simple and is not normally an accurate reflection of the various stages of a true enzyme assisted reaction. A more detailed model was later proposed by Briggs and Haldane, where the complex has two phases (ES and EP). This has the same form as Eqs. 1.17 and 1.18, but again with a different constant. We can obviously extend the analysis to as complex a model as we wish (and we will look at some of these below), but the analysis does become increasingly laborious to perform. In practice, quite often a simple approximation is sufficient to capture most of the behaviour of the system, especially if it is operating near either a linear or a saturated regime.

1.3.3 Enzyme Cooperativity

Some enzymes can bind more than one substrate molecule, such that the binding of one substrate molecule affects the binding of subsequent molecules. This is known as **cooperativity** and is involved in one of the most important bindings found in the human body: that of oxygen to haemoglobin in the blood. Although the analysis is quite complicated, the final result is relatively simple: if n substrate molecules can bind to the enzyme, the rate of reaction is given by:

$$V = \frac{V_{max}s^n}{K^n + s^n} \tag{1.19}$$

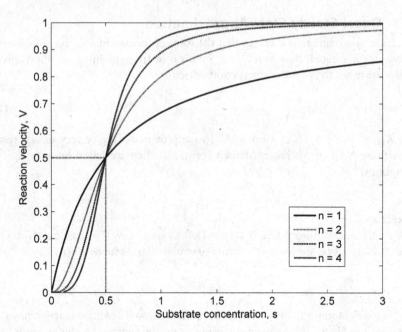

This is known as the **Hill equation**. It is frequently used as an approximation for reactions where the intermediate steps are not well known and is then derived from fits to experimental data. The shape of the Hill equation for different values of n is shown in ☐ Fig. 1.8, illustrating how this can be used to fit different shapes of curves derived from experimental data. In particular, note that the curve becomes much sharper for larger values of n (the asymptote is of course the same value in all cases).

This relationship is particularly useful in the case of oxygen binding to haemoglobin, where the reaction equation:

$$Hb + 4O_2 \underset{k_-}{\overset{k_+}{\rightleftharpoons}} Hb(O_2)_4 \tag{1.20}$$

turns out to imply a fractional filling of available haemoglobin sites, S, of:

$$S = \frac{[O_2]^n}{K^n + [O_2]^n} \tag{1.21}$$

with $n = 4$. In fact, a much better fit to experimental data is found with $n = 2.5$, implying that the later sites prefer to fill up if the early sites are already full. This is known as positive cooperativity, although the process is not yet completely understood. This example, which is known as the oxygen saturation curve and which is very important in respiration, as we will see in ▶ Chap. 9, provides a useful early illustration of a model that provides a good understanding of how the reaction processes occur, but where the result has to be adapted to mimic the real behaviour. It is worth noting that the oxygen saturation curve is also affected by pH, temperature and CO_2 levels, amongst other things, all of which mean that the curve has to be adapted to fit experimental data.

Exercise E
a. Show that the Hill equation can be written as a straight-line plot if the value of V_{max} is known.
b. Sketch this function and explain how the intercepts with the two axes can be used to calculate the values of n and K_m. Given the values in the table below, calculate the values of n and K_m, assuming that $V_{max} = 1$ mM. Is the Hill equation a good fit to this data set?

Table E.1

Substrate conc. (mM)	0.2	0.5	1.0	1.5	2.0	2.5	3.5	4.5
Reaction velocity (mM/s)	0.01	0.06	0.27	0.5	0.67	0.78	0.89	0.94

1.3.4 Enzyme Inhibition

In many cases, it is very important to have **control** over the reaction rate, so that it can be altered under different conditions, for example speeded up when demand for the product is greater or slowed down when demand drops. An **enzyme inhibitor** is thus often used to reduce the rate of reaction, as explained below. There are two types that we will look at here: **competitive** and **allosteric** inhibitors.

1.3.4.1 Competitive Inhibitors

In this case, the inhibitor species combines with the enzyme to form a compound, which essentially removes some of the enzyme from the system, preventing it from forming the product: hence the idea of competition for the enzyme. The reactions are thus:

$$E + S \underset{k_{-1}}{\overset{k_{+1}}{\rightleftharpoons}} C_1 \overset{k_{+2}}{\rightarrow} E + P \tag{1.22}$$

$$E + I \underset{k_{-3}}{\overset{k_{+3}}{\rightleftharpoons}} C_2 \tag{1.23}$$

There are now six differential equations:

$$\frac{ds}{dt} = k_{-1}c_1 - k_{+1}se \tag{1.24}$$

$$\frac{de}{dt} = k_{-1}c_1 + k_{+2}c_1 - k_{+1}se + k_{-3}c_2 - k_{+3}ie \tag{1.25}$$

$$\frac{dc_1}{dt} = k_{+1}se - (k_{+2} + k_{-1})c_1 \tag{1.26}$$

$$\frac{dp}{dt} = k_{+2}c_1 \tag{1.27}$$

$$\frac{di}{dt} = k_{-3}c_2 - k_{+3}ie \tag{1.28}$$

$$\frac{dc_2}{dt} = k_{+3}ie - k_{-3}c_2 \tag{1.29}$$

Thankfully, this isn't as complicated as it looks. The normal assumption made in the analysis is that both of the compounds are in quasi-steady-state. We can also see that if we add the differentials for the enzyme and the two compounds, they add up to zero (so these three variables always add up to a constant, which we will again call e_o).

The rate of formation of the product is then found to be:

$$V = \frac{dp}{dt} = k_{+2}c_1 = \frac{V_{max}s}{K_m(1 + i/K_i) + s} \tag{1.30}$$

where $K_i = k_{-3}/k_{+3}$. Note that if i is set to zero, we get back to the original Michaelis-Menten equation, just as we would expect.

Exercise F
a. Derive the result in (1.30), given that $dc_1/dt = 0$ and $dc_2/dt = 0$ in quasi-steady-state, and that $c_1 + c_2 + e = e_o$.
b. Sketch the result for different values of inhibitor. Explain how the inhibitor affects the intercepts on the $1/V$ versus $1/s$ plot.

It is worth noting at this point that you will often see multiple reaction equations combined into a single schematic form. The example above can be re-written in the form below (where we have omitted the rate constants for simplicity).

$$E \;\; + \;\; S \;\; \rightleftarrows \;\; ES \;\; \rightarrow \;\; E \;\; + \;\; P$$

$$+$$

$$I$$

$$\updownarrow$$

$$EI$$

This diagrammatic form can be very helpful in seeing how the reaction equations link together and understanding how substances 'move' around the system, although we can of course always write them in the original form.

1.3.4.2 Allosteric Inhibitors

In this case the inhibitor binds to the enzyme in such a way as to prevent the product being formed. For example, the inhibitor might bind to a different site on the enzyme preventing the complex converting into the product. This can be modelled as the inhibitor binding to the complex, written in schematic form as shown below (note how we have moved to simply sketching the diagram to explain the system before writing down the equations).

$$E \; + \; S \; \rightleftarrows \; ES \; \rightarrow \; E \; + \; P$$

$$+$$

$$I$$

$$\updownarrow$$

$$ESI$$

A schematic of allosteric inhibition is shown in ◘ Fig. 1.9, alongside allosteric activation. The latter is essentially the complementary process, where an activator is required to enhance the role of the enzyme: like the enzyme, it is not used up in the reaction, but its presence affects the overall reaction rate.

The rate of formation of product is now:

$$V = \frac{V_{max}s}{K_m + (1 + i/K_i)s} \tag{1.31}$$

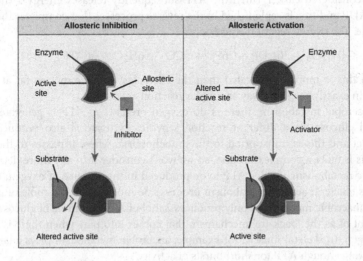

◘ **Fig. 1.9** Allosteric inhibition and activation (this figure is taken, without changes, from OpenStax College under license: ► http://creativecommons.org/licenses/by/3.0/)

1

This is quite similar to the previous equation, but you should be able to spot the difference in its behaviour and how the rate of formation is altered in different ways by inhibition or activation.

In practice this result is only true for an allosteric inhibitor that is **uncompetitive**, i.e. one that doesn't bind to E to form EI. It is of course possible to have inhibitors that are both allosteric and competitive in which case the analysis becomes more complex. We won't go into these, as there are more detailed treatments of these elsewhere. However, you have by now at least got an idea of how we write down reaction equations and how we analyse them. However complex the system of equations, it will be analysed in the same way.

1.3.5 Cellular Metabolism

We will briefly consider the processes that govern the metabolism of the cell here, as these are governed by a number of reaction equations. As we have already noted, ATP is the energy source used by the cell. Cellular metabolism is based on the conversion of glucose into other products with ATP being generated. Glucose travels from the bloodstream into the surrounding tissue due to a concentration gradient (something we will meet in ▶ Chap. 4). It is then trapped in the cells through the action of the enzyme hexokinase that adds a phosphate to a glucose molecule to convert it into glucose-6-phosphate, such that it cannot easily cross the cell membrane. This molecule is then converted into various other molecules in the metabolic process.

There are two types of metabolism: **aerobic** and **anaerobic**. Oxygen is key to the first of these with ATP being formed from ADP and glucose (the reverse process is what we considered earlier when the ATP subsequently releases energy). There is a store of ADP and this is replenished in the reverse processes that completes the metabolic cycle. The overall reaction equation is:

$$6O_2 + C_6H_{12}O_6 + 36ADP + 36P_i \rightarrow 6CO_2 + 6H_2O + 36ATP \tag{1.32}$$

Although this is more complicated than the examples we have met so far, it can be analysed in exactly the same way as simpler reaction equations.

In anaerobic metabolism, there is no oxygen present, so ATP is generated from ADP and glucose via a different reaction scheme. Pyruvate is also produced as a by-product and this is transported to the mitochondria, where it passes to the Krebs cycle. This is quite a complex process, so we won't consider it in any more detail here. Anaerobic metabolism enables ATP to be produced in the absence of oxygen, but it is much less efficient: aerobic metabolism produces 36 mol of ATP per mole of glucose, whereas anaerobic metabolism only produces 2 mol of ATP per mole of glucose. It can be thought of as the 'back-up' mechanism that comes into play when there is insufficient oxygen. In skeletal muscle, for example, anaerobic metabolism plays a large part in generating enough ATP for short bursts of activity.

1.4 Conclusions

In this chapter we have looked at how human cells are constructed, what substances they are made of and considered how to analyse simple chemical processes using reaction equations. You should now be familiar with the basic make-up of cells and be able to understand how to write down any system of reactions, using both diagrammatic form and reaction equations, and to do some basic analysis.

Cancer: an example of abnormal cell structure and function

We have considered how the cell is constructed and we will shortly consider how it operates. Of course, this can go wrong, and **cancer** is the result of the abnormal growth of cells due to DNA mutation or damage. Such damage can be caused by many factors and is amplified by activities such as smoking. In normal, healthy tissue, cells divide in a tightly regulated manner with processes both stimulating and inhibiting the process. However, in cancer cells, cell division becomes unregulated and a single cell that can continue to divide itself and to survive can produce enough cells to kill a person. Within the clusters of cells that form a tumour anaerobic respiration often becomes dominant. This gives the tumour an advantage where there is an insufficiency in the rate at which oxygen can be supplied to the tumour, arising from its fast growth outpacing the rate at which new blood vessels grow to supply the oxygen demand.

Cancerous tumours are difficult enough to treat due to the fact that they appear to the body a lot like normal cells. But, treatment would be possible through highly targeted therapy were it not for the fact that cancer cells can evolve. As generations of cancer cells are generated, small genetic differences appear. Thus, when treatment for cancer kills some specific cancer cells, the fittest cells survive and continue to reproduce. This makes cancer treatment so difficult and explains why the modern practice is to prescribe combinations of drugs, in order to try to kill off all the cells before evolution can occur. Thankfully survival rates for many cancers have now improved drastically, but cancer remains one of the biggest killers around the world.

Cellular Homeostasis and Membrane Potential

Contents

© Springer Nature Switzerland AG 2020
M. Chappell and S. Payne, *Physiology for Engineers*, Biosystems & Biorobotics 24,
https://doi.org/10.1007/978-3-030-39705-0_2

In ▶ Chap. 1 we considered the structure of the human cell. We will now consider what it is made of in more detail and in particular think about how this determines how it operates. The relationship between **structure** and **function** is a very common idea in physiological systems and one that we will return to again later as we consider more complex systems in detail.

2.1 Membrane Structure and Composition

The human cell can be considered to consist of a bag of fluid with a wall that separates the internal, or intracellular, fluid (ICF) from the external, or extracellular, fluid (ECF): this wall is termed the plasma **membrane**. The membrane consists of a sheet of lipids two molecules thick: lipids being molecules that are not soluble in water but are soluble in oil. The cellular lipids are primarily phospholipids, as illustrated in ◘ Fig. 2.1, they have one end that is hydrophilic and one that is hydrophobic (are attracted to and repelled by water molecules respectively). The hydrophobic ends tend to point towards each other, and away from the aqueous environments inside and outside the cell, hence the two-molecule thickness of the membrane.

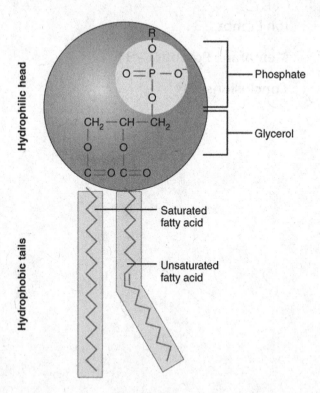

◘ Fig. 2.1   Structure of a phospholipid with both hydrophobic and hydrophilic ends (this figure is taken, without changes, from OpenStax College under license: ▶ http://creativecommons.org/licenses/by/3.0/)

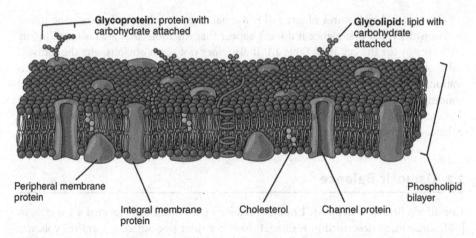

Fig. 2.2 Schematic of membrane structure (this figure is taken, without changes, from OpenStax College under license: ▶ http://creativecommons.org/licenses/by/3.0/)

Table 2.1 Compositions of intracellular and extracellular fluids for a typical cell

	Internal concentration (mM)	External concentration (mM)
K^+	125	5
Na^+	12	120
Cl^-	5	125
P^{-*}	108	0
H_2O	55,000	55,000

*P refers to non-ionic substances, largely proteins and amino acids, inside the cell that are charged. The overall net charge is negative

Substances can cross the membrane if they can dissolve in the lipids, which will not be true of most of the species that are in aqueous solution, such as ions. However, some electrically charged substances, which cannot readily pass through the lipid sheets, do cross the membrane. This is because the membrane is full of various types of protein molecules, some of which bridge the lipid layer. Sometimes these form pores or channels through which molecules can pass. ❏ Figure 2.2 shows a schematic of the membrane structure that illustrates these ideas.

Outside the cell, in the ECF, the main positively-charged ion is **sodium** (Na^+) with a small amount of **potassium** (K^+) and **chloride** (Cl^-) ions, as shown in ❏ Table 2.1. Note that the relative quantities are reversed in the ICF. The balance of charge is provided inside the cell by a class of molecules that include protein molecules and acidic amino acids (likewise outside the cell, but these will be ignored here for reasons that we will see later).

2

One of the key features of any cell is the balance of molecules between the inside and outside, yet at first glance it doesn't appear that the molecules are balanced in any obvious way for the cell in ◘ Table 2.1. It also does not seem obvious why the individual molecules do not diffuse in and out of the cell such that the ICF and ECF concentrations are equal. In general we are interested in how the cell maintains its internal conditions despite what is going on outside: the property called **homeostasis**. If we want to understand how the cell maintains this imbalance we first need to consider the balance of cell volume.

2.2 Osmotic Balance

Consider a litre of water with 1 mol of dissolved particles: this is termed a 1 molar, or 1 M, solution, as described in ▶ Chap. 1. Now consider two adjacent identical volumes with different molarities (say 100 mM and 200 mM) and a barrier between them. If the barrier allows both water and the solute to pass, equilibrium will be reached with equal levels of the solute (150 mM) and the barrier will not move, as might be expected.

However, if the barrier allows only water to cross, enough water will have to cross the barrier for the concentrations to balance and the barrier will thus move. Equilibrium will then be reached with the same concentrations (150 mM) but different volumes, 2/3 and 4/3 L, as illustrated in ◘ Fig. 2.3. The volumes are calculated by remembering that the concentrations must balance and that the number of moles of the solute cannot change. This process can be thought analogous to pressurised chambers with initial pressures (equivalent to the concentrations) and volumes: the barrier is like a piston, moving until pressure equilibrium is reached.

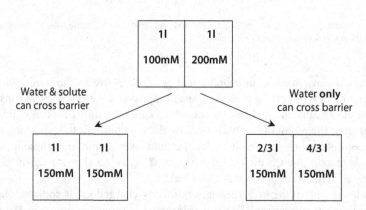

◘ **Fig. 2.3** Example of balance of concentration with solute able to cross (left) and with solute unable to cross (right)

Now consider a slightly different example with a cell with an intracellular concentration of a substance P and an extracellular concentration of a substance Q: neither P nor Q is able to cross the membrane. There are three possibilities:

1. The concentration of Q is equal to that of P (**isotonic**): cell volume remains constant.
2. The concentration of Q is greater than that of P (**hypertonic** solution): cell volume decreases, and the cell ultimately collapses if the difference in concentration is sufficiently large.
3. The concentration of Q is less than that of P (**hypotonic** solution): cell volume increases eventually rupturing the membrane if the concentration difference is sufficiently large.

A hypotonic solution is defined as one that makes the cell increase in size and a hypertonic solution is one that makes the cell decrease in size. An illustration of the three types of behaviour is shown in ◘ Fig. 2.4.

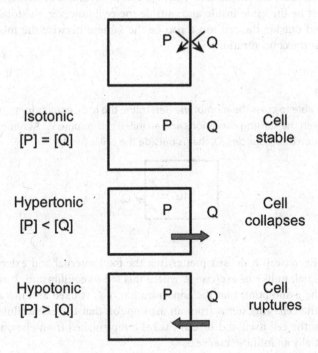

◘ Fig. 2.4 Cell behaviour in solutions of different concentration. Since neither P nor Q are able to cross the cell membrane, any adjustments in concentration can only occur by movement of water, i.e. osmosis

2

Exercise A
For the following case determine the final, equilibrium, cell volume given the starting conditions and an initial cell volume V_0.

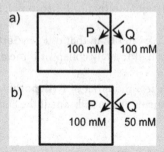

So far we have only considered what happens if water can cross the membrane. If Q (or P) is able to cross the membrane, then the cell behaves differently. The concentration of Q must be the same inside and outside the cell; however, the total concentration inside and outside the cell must also be the same otherwise the membrane will move to adjust the concentrations.

Exercise B
If Q is now able to cross the membrane determine the final, equilibrium, cell volume given the starting conditions and an initial cell volume V_0. Assume that the concentration of the species, Q, that is outside the cell is fixed.

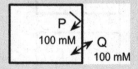

Note that in Exercise B it doesn't matter that the total internal and external concentrations are equal, unlike in exercise A where this led to equilibrium. It is also worth questioning the assumption that the concentration of Q is fixed and not altered by Q moving into the cell. This stems from an assumption that the extracellular space is a lot larger than the cell itself and/or can easily be replenished from elsewhere, i.e. that there is essentially an infinite reserve of Q.

The requirement for the total concentration to be equal inside and outside the cell is actually very simply a requirement that the concentration of water balances. It might seem strange to talk about water concentration since there is so much of it compared to the other species, but **osmosis** is essentially the process of balancing water concentration.

The total concentration is often referred to as the **osmolarity**: the higher the osmolarity the lower the concentration of water and vice versa. A solution containing 0.1 M glucose and 0.1 M urea would have a total concentration of non-water species of 0.2 M and thus an osmolarity of 0.2 Osm. Care does need to be taken with solutions of substances that dissociate, however: for example, a 0.1 M solution of NaCl is a 0.2 Osm solution since you get both free Na^+ and Cl^-. In practice the osmolarity could be lower than this if the ions in solution interacted, but this is not common in biological systems.

Exercise C

For the following cases determine the final, equilibrium, cell volume given the starting conditions and an initial cell volume V_0.

a)

	P ⤬ Q	
	100 mM	100 mM
	R ⟷ R	
	100 mM	100 mM

b)

	P ⤬ Q	
	100 mM	50 mM
	R ⟷ R	
	100 mM	100 mM

We have considered how the cell might maintain its volume despite imbalances in the concentrations of species inside and outside the cell. In fact we have met our first principle, the **principle of concentration balance**.

2.3 Conservation of Charge

Now consider the slightly more complex example in ◻ Fig. 2.5, which is a very basic model v a cell. Inside the cell are found organic molecules, P, which cannot pass through the barrier. The internal Na^+ is also trapped, whereas Cl^- can pass freely through the barrier. The concentrations of P and Na^+ inside the cell are 100 and 50 mM respectively.

To analyse this model, there are two quantities that must be in balance: charge and concentration. The fact that the positive and negative charges must balance within any compartment is called the **principle of electrical neutrality**, which states that the bulk concentration of positively charged ions must equal the bulk concentration of

◻ **Fig. 2.5** Cell model example

50 mM Na⁺ ⤬ Na⁺	b
a Cl⁻ ⟷ Cl⁻	c
100 mM P ⤬	

negatively charged ions. Essentially this is due to the fact that under biological conditions, so few positively and negatively charged ions have to move to generate any membrane potential (which we will meet later) that we can assume that they balance at all times.

From charge balance:

$$a = 50 \tag{2.1}$$

$$b = c \tag{2.2}$$

From total concentration balance:

$$50 + a + 100 = b + c \tag{2.3}$$

Hence:

$$b = c = 100 \tag{2.4}$$

Note that, unlike in the previous section, the concentrations of Cl^- are not equal inside and outside the cell: this is due to the influence of the charge balance. In fact at this point we can more carefully define the principle of concentration balance to only refer to the concentrations of **uncharged** species. For a simple cell model the only uncharged species that can freely cross the membrane to any significant degree is water, thus the principle of concentration balance might be called the principle of osmotic balance.

2.4 Equilibrium Potential

So far we have only considered concentration equilibrium: however, there is another important factor that drives ions across a cell membrane. In addition to the concentration gradient that drives ions from a region of high concentration to a region of low concentration, there is an **electrical** potential difference across the membrane.

For the membrane shown in ❏ Fig. 2.6, the difference in voltage between the inside and the outside of the cell is given by the Nernst equation:

$$E_X = V_{in} - V_{out} = \frac{RT}{ZF} \ln\left(\frac{[X]_{out}}{[X]_{in}}\right), \tag{2.5}$$

where R is the gas constant, T is absolute temperature (in Kelvin), Z is the valence of the ion and F is Faraday's constant (96,500 coulombs/mol_univalent_ion). The quantity in the equation above is known as the **equilibrium potential** and only applies for a

❏ **Fig. 2.6** Membrane and membrane potential

$[X]_{in}$ $[X]_{out}$

$E_m = V_{in} - V_{out}$

V_{in} V_{out}

single ion that can cross the barrier. At standard room temperature, the equation can be re-written as:

$$E_X = \frac{58 \text{ mV}}{Z} \log_{10}\left(\frac{[X]_{out}}{[X]_{in}}\right), \tag{2.6}$$

where we have changed from a natural logarithm to a base-10 logarithm purely for convenience.

There can, however, obviously only be a single potential across the membrane, the **membrane potential**, thus if there are two ions that can cross the membrane (in the real cell these are K^+ and Cl^-), then the equilibrium potential must be the same for both. Hence:

$$E_m = 58 \text{ mV} \log_{10}\left(\frac{[K^+]_{out}}{[K^+]_{in}}\right) = -58 \text{ mV} \log_{10}\left(\frac{[Cl^-]_{out}}{[Cl^-]_{in}}\right), \tag{2.7}$$

which on re-arranging becomes:

$$\frac{[K^+]_{out}}{[K^+]_{in}} = \frac{[Cl^-]_{in}}{[Cl^-]_{out}}. \tag{2.8}$$

This is known as the **Donnan** or **Gibbs-Donnan equilibrium equation.**

Exercise D

Suppose that two compartments, each of one litre in volume, are connected by a membrane that is permeable to both K^+ and Cl^-, but not permeable to water or the protein X. Suppose further that the compartment on the left initially contains 300 mM K^+ and 300 mM Cl^-, while the compartment on the right initially contains 200 mM protein, with valence -2, and 400 mM K^+.

(a) Is the starting configuration electrically and osmotically balanced?

(b) Find the concentrations at equilibrium.

(c) Why is [K^+] in the right compartment at equilibrium greater than its starting value, even though [K^+] in the right compartment was greater than [K^+] in the left compartment initially? Why does K^+ not diffuse from right to left to equalize the concentrations?

(d) What is the equilibrium potential difference?

Exercise D is an interesting example of the counterintuitive differences in ion concentration that can arise due to the balance of electrical configuration. However, it is not like our cell model because we had only a fixed total amount of each ion available and the whole system was thus closed.

◘ Fig. 2.7   Simple cell model

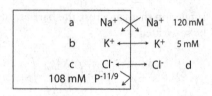

2.5 A Simple Cell Model

We started out by trying to understand cell homeostasis and how an imbalance in concentrations of ions inside and outside of the cell could be maintained. We now have three principles to apply when we analyse cell concentrations:
1. Concentration (osmotic) balance.
2. Electrical neutrality.
3. Gibbs-Donnan equilibrium.

We are now ready to try and build a simple model of a cell at equilibrium, ◘ Fig. 2.7. Inside the cell are found Na^+, K^+ and Cl^- as well as some negatively charged particles, termed P, that represent an array of different molecules, including proteins. Outside the cell are found Na^+, K^+ and Cl^- ions, where K^+ and Cl^- are free to cross the membrane.

> **Exercise E**
> ◘ Figure 2.5 is incomplete since some of the concentrations have not been given.
> (a) Write down and solve the appropriate equations to calculate the unknown concentrations in the figure. Note that the charge of P is −11/9 (about −1.22)[1], which means that the charge equilibrium equation must be written down carefully.
> (b) What is the value of the membrane potential in this example?
> (e) If the cell membrane were permeable to sodium, calculate the equilibrium potential for sodium. What would then happen to the cell?

This exercise takes us close to a realistic model of the cell: you should find that the values of concentration that you get are the same as ◘ Table 2.1. Note that it will remain in this state indefinitely without the expending of any metabolic energy: a very efficient structure. However, you will have found that if the cell is permeable to Na^+, its equilibrium potential is very different from the membrane potential that you calculated when the membrane is impermeable to Na^+. Thus if the membrane were permeable to sodium then it would be impossible to achieve equilibrium due to the proteins also present in the cell: the cell would simply grow until rupture.

---

1   Some other texts round down to a charge of −1.2, which appears to be very close to that here, but will result in quite different concentrations for some of the ions if you try to use it.

2.6 Ion Pumps

Unfortunately, the real cell actually does expend metabolic energy in order to remain at equilibrium. The reason for this is that the cell wall is actually slightly permeable to Na^+, which implies that our model cell will not remain in an equilibrium state. The answer to this problem is that there is something called a sodium pump, which we will now examine briefly.

An **ion pump** is a mechanism that absorbs energy to move ions against a concentration or electrical gradient, rather like a heat pump. The ion pump gets its energy from the ATP that we met in ▶ Chap. 1. For Na^+, as fast as it leaks in due to the concentration and electrical gradients, it is pumped out. Na^+ thus effectively acts as if it cannot cross the membrane, but the cell is now a **steady state**, requiring energy, rather than an equilibrium state, which would require no energy.

The common symbol for the pump is shown in ◘ Fig. 2.8, which also shows that the pump needs K^+ ions outside the cell to pump inside in return for Na^+ ions inside. The protein on the cell outer surface needs K^+ to bind to it before the protein can return to a state in which it can bind another ATP and Na^+ ions at the inner surface. Since the K^+ ions bound on the outside are then released on the inside, the pump essentially swaps Na^+ and K^+ ions across the membrane and is thus more correctly known as the Na^+/K^+ pump and the membrane-associated enzyme as a Na^+/K^+ ATPase. We will revisit the ion pump in more detail in ▶ Chap. 4.

2.7 Membrane Potential

Now that we have reconsidered the cell as a steady state system (rather than an equilibrium system), we need to reconsider the membrane potential. Previously in Exercise E we had $E_m = E_K = E_{Cl} = -81$ mV. However, now that we also have a contribution from Na^+ with $E_{Na} = +58$ mV, the membrane potential will have to settle somewhere between these extremes. This actually depends upon both the ionic concentrations and the membrane **permeability** to the different ions. Clearly if the permeability to a particular ion is zero, it contributes nothing to the potential, whereas with a high permeability it contributes significantly more. The permeability of the membrane to different ions is absolutely vital in our understanding of the operation of the cell.

The permeability of a membrane to a particular ion is simply a measure of how easily those ions can cross the membrane. In electrical terms, it is equivalent to the inverse of resistance (i.e. conductance). We will consider why the permeabilities are different for different ions in ▶ Chap. 3, but for now, we will note that the permeability is related to the number of channels that allow the ions to pass through and the ease of passage through the channels.

◘ **Fig. 2.8** Schematic of sodium/potassium pump

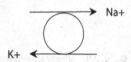

The relationship between membrane potential and the concentrations and permeabilities of the different ions in the cell is known as the **Goldman equation:**

$$E_m = 58 \text{ mV} \log_{10} \left(\frac{p_K \left[K^+\right]_o + p_{Na} \left[Na^+\right]_o + p_{Cl} \left[Cl^-\right]_i}{p_K \left[K^+\right]_i + p_{Na} \left[Na^+\right]_i + p_{Cl} \left[Cl^-\right]_o} \right), \tag{2.9}$$

where p denotes permeability. Note that because Cl^- has a negative valence the inner and outer concentrations are the opposite way round to those for Na^+ and K^+. For a membrane that is permeable to only one ion, the Goldman equation reduces immediately to the Nernst equation.

In practice, the contribution of Cl^- is negligible and hence the equation is usually encountered in the form:

$$E_m = 58 \text{ mV} \log_{10} \left(\frac{\left[K^+\right]_o + b \left[Na^+\right]_o}{\left[K^+\right]_i + b \left[Na^+\right]_i} \right), \tag{2.10}$$

where in the resting state $b = p_{Na}/p_K$ is approximately 0.02. For the typical resting state with the concentrations given previously, the membrane potential is approximately -71 mV. The membrane potential is closer to the value for K^+, since the permeability to K^+ is much greater than that for Na^+. However, changes in the relative permeability can produce large changes in the membrane potential between these two values.

Since the membrane potential is equal to neither the values for Na^+ nor for K^+, there is a leakage of both K^+ out of and Na^+ into the cell: hence the role of the Na^+/K^+ pump to maintain the membrane potential at a steady state value. A more complete model for cell is shown in ◘ Fig. 2.9 that also includes the forces acting on the ions. Note that the net charge on the inside of the cell is negative therefore the electrostatic forces acting on both Na^+ and K^+ in inward, whereas on Cl^- it is outward. It is easy to see why at the very least a pump for Na^+ is required.

Although we have ignored Cl^- in the calculation of the membrane potential, it is affected by it: the equilibrium membrane potential for Cl^- is -80 mV, so either the concentration will change (as in some cells) or a Cl^- pump is used to maintain a steady state level of Cl^-. Less is known about this pump than the Na^+/K^+ pump.

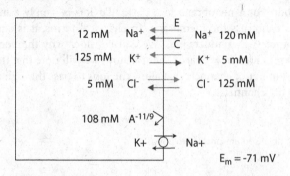

◘ **Fig. 2.9**   Steady state cell model, showing the electrical (E) and chemical (C) forces acting on the ions

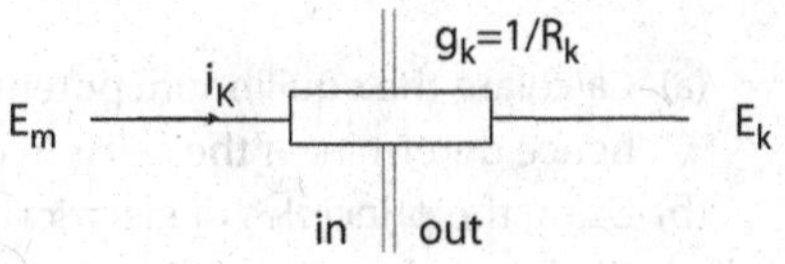

Fig. 2.10 Modelling the cell membrane as a resistance/conductance for each ion

Since a difference in membrane potential from the equilibrium value for an individual ion causes a movement of ions across the membrane we can introduce a new concept, that of membrane conductance, as defined by:

$$i_K = g_K(E_m - E_K), \tag{2.11}$$

$$i_{Na} = g_{Na}(E_m - E_{Na}), \tag{2.12}$$

$$i_{Cl} = g_{Cl}(E_m - E_{Cl}). \tag{2.13}$$

Since $E_m = -71$ mV, $E_K = 80$ mV and $E_{Na} = 58$ mV from above, the potassium current is positive and the sodium current is negative. By convention, an outward current is positive and an inward current is negative. In the steady state the net current is zero, which is the basis of the Goldman equation. The conductance is related to both the permeability and the number of available ions in the solution. Note that conductance is the inverse of resistance and so in electrical terms, the membrane can be considered as a resistor as shown in **Fig. 2.10**.

The meaning of permeability can be explored in more detail by noting that the membrane is full of protein channels that permit different ions to pass through. These channels can be considered to be controlled by a gate that is either open or closed (this mechanism is known as channel gating). Although the channels are slightly more complicated than this, it is a valid first approximation: we will examine this in more detail in ▶ Chap. 3. Rather like an electrical switch, each channel is thus either 'on' or 'off' as far as current is concerned. Since there are a very large number of channels, the permeability of the membrane can be controlled to a high degree of accuracy by the opening of different numbers of channels. This ability to change the membrane permeabilities is a major factor in the behaviour of cells and this will be examined in ▶ Chap. 3 when we consider the action potential.

Exercise F

A simple model for a cardiac myocyte (heart muscle cell) can be built using the concentrations of the ions to which the membrane is permeable given in the table.

Ion	Internal concentration (mM)	External concentration (mM)
Na^+	10	145
K^+	140	4
Cl^-	30	114
Ca^{2+}	10^{-4}	1.2

2

(a) Calculate the equilibrium potentials for all the ions in the cardiac myocyte and hence determine if the cell is in equilibrium.

(b) Using the principles of electrical neutrality and osmotic balance determine the internal concentration and overall charge of other charged species (e.g. proteins) within the cell that are unable to cross the cell membrane, assuming zero external concentration of any other species.

(c) Show that this cardiac myocyte cell is not in a steady state.

This final exercise explores a cell with a specific function that we will meet in ▶ Chap. 7. Whilst a lot of the values for ionic concentrations, charges and potentials are not significantly different from the model cell we have considered in this chapter, this cell is neither in equilibrium nor even in steady state. We might have been wrong to ignore any other charged species outside the cell. Otherwise, like the simple cell model, we might worry that this cell would expand until rupture. Note that ion pumps wouldn't help here as the concentrations we have do not satisfy all the principles and thus the cell cannot be in steady state; the ion pumps only helped to maintain the steady state in the simple cell model. It turns out that cardiac myocytes never reach a steady state and the concentrations fluctuate over a cycle, these values (probably) just representing the 'resting' state. We will return to this in ▶ Chap. 7.

2.8 Conclusions

In this chapter we have considered how the cell can maintain homeostasis despite differences in concentration of ions and other charged species both inside and outside of the cell membrane. You should now understand the principles of concentration balance, electrical neutrality and Gibbs-Donnan equilibrium and be able to apply them to a cell model. You should also be able to calculate equilibrium potentials for individual ions as well as the membrane potential and understand why these are not always the same value. Finally you should now appreciate why cells use energy to maintain homeostasis.

The Action Potential

Contents

© Springer Nature Switzerland AG 2020
M. Chappell and S. Payne, *Physiology for Engineers*, Biosystems & Biorobotics 24,
https://doi.org/10.1007/978-3-030-39705-0_3

In ▶ Chaps. 1 and 2 we considered properties of cells that are universally found throughout the body. We now begin to consider more specific properties found in certain types of cell that are vital for the correct functioning of the human body. In particular, we will focus here on 'excitable' cells, which can generate active electrical responses that act as signals for other events. Whereas in ▶ Chap. 2 the cells were in a steady state, now we will meet cells that use transient changes in ion permeability and membrane potential as a means to generate signals. Specifically, we are going to consider the basic signal of the nervous system: the action potential.

3.1 Na$^+$/K$^+$ Action Potential

As we saw in ▶ Chap. 2, it is the relative permeability of the membrane to Na$^+$ and K$^+$ that determines the membrane potential.

$$E_m = 58\,\text{mV}\ \log_{10}\left(\frac{[\text{K}^+]_o + b[\text{Na}^+]_o}{[\text{K}^+]_i + b[\text{Na}^+]_i}\right). \tag{3.1}$$

In the resting state, the ratio $b = 0.02$, as given in ▶ Chap. 2, and thus $E_m = -71$ mV. But if the membrane permeability to Na$^+$ were suddenly increased by a significant factor, this ratio would increase and the membrane potential swing from close to E_K (−80 mV) to close to E_{Na} (+58 mV). This is essentially all that is required to generate the **action potential**, which is a transient change in the membrane potential of the cell. The action potential arises because the gates which determine the permeability of the membrane are themselves controlled by the membrane potential. The ion channels found in nerve cells are called **voltage-gated** specifically because they respond to changes in the membrane potential by either opening or closing.

> **Exercise A**
> Calculate the membrane potential if b were to increase by three orders of magnitude to 20, assuming that the ionic concentrations are still at the steady state values in ◘ Table 2.1. What would be the implications for the ionic concentrations of this change?

Since E_m is a logarithmic function of the relative permeability it takes orders of magnitude changes in b to make a substantial difference. You will have noticed in Exercise A that an increase in relative permeability either implies that the permeability to K$^+$ has reduced or to Na$^+$ has increased, in practice though it is the latter. A change in permeability will lead to a flow of ions driven by the different membrane potential, the cell will no longer be in steady state.

A typical action potential is shown in ◘ Fig. 3.1, which would correspond to a nerve cell: we will see examples of variants on this later. Notice that it is a transient process, as ultimately the membrane potential settles back to steady state. We might describe it as a voltage 'spike', for a nerve cell this will last of the order of a millisecond.

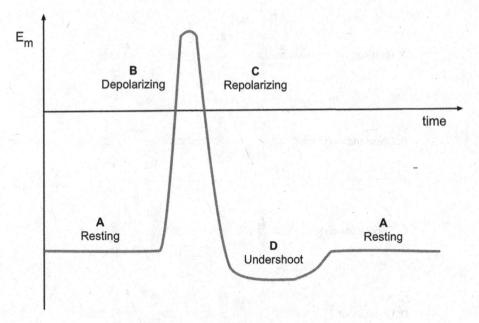

● **Fig. 3.1** The membrane potential during a typical (nerve cell) action potential

The process of the action potential follows these steps, summarised in ● Fig. 3.2 and ● Table 3.1:

A. Resting state: The number of Na^+ channels that are open is small and $b = p_{Na}/p_K \sim 0.02$.

B. Depolarization: If the membrane potential becomes more positive then more Na^+ channels open, thus b will increase and E_m will become even more positive. This positive feedback loop means that there is a large and rapid swing in E_m, meaning that the action potential continues irrespective of any further external influence. Thus the action potential is known as an 'all or nothing' event.

C. Repolarization: The Na^+ channel actually has two gates: m that is normally closed and h that is normally open. Upon depolarization m opens rapidly, but h closes much more slowly and it is this timing difference that leaves time for depolarization before repolarization kicks in. Additionally, there are voltage sensitive K^+ channels with n gates that are normally closed that also respond slowly to the depolarization. Thus p_K also increases on depolarization and like the h gate this drives E_m back toward the resting value.

D. Undershoot: Due to the n gates p_K is greater than usual so that b ends up being smaller than the resting value, which causes E_m to undershoot until the n gates eventually return to normal.

E. Refractory period: Whilst the membrane potential may have returned to the resting state the h gates will initially still be shut blocking the Na^+ channels even if the m gates were to reopen. Thus there is a period in which a new action potential cannot be generated. In practice the undershoot and refractory periods

3

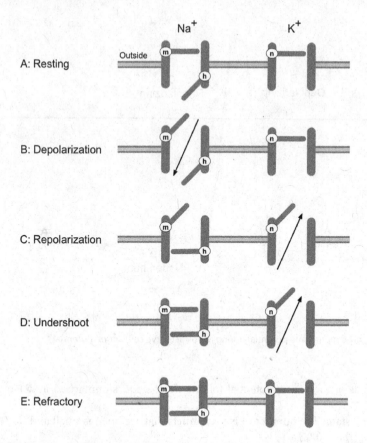

Na⁺   **K⁺**

A: Resting

B: Depolarization

C: Repolarization

D: Undershoot

E: Refractory

◘ Fig. 3.2   Schematic of voltage-sensitive channels during action potential

◘ Table 3.1   Overview of gate behaviour during an AP

Phase	Na⁺ gates		K⁺ gate	b
	m	h	n	
A: Resting state	Closed	Open	Closed	~0.02
B: Depolarising	Opens quickly	Closes slowly	Opens slowly	0.02 → 20
C: Repolarising	Open			20 → ~0.02
D: Undershoot	Closing quickly	Closed	Open	<0.02
E: Refractory	Closed	Opening slowly	Closing slowly	<0.02 → ~ 0.02

overlap and the extent to which it is raised K^+ permeability or inability for an increase Na^+ permeability that prevents another AP from firing depends upon the precise timings of the gates.

In practice the action potential only occurs once a certain threshold of membrane potential is reached, usually around 10–20 mV above its resting value. For smaller depolarization the efflux of K⁺ induced by the change in E_m exceeds the efflux of Na⁺, leading to a negative feedback process that suppresses further depolarization. The actual value of this threshold depends upon a variety of factors, particularly the packing density of the Na⁺ channels and the relationship between the membrane potential and the opening pattern. Some neurons are highly sensitive, whilst others require a very large depolarization to stimulate a potential.

3.2 Ca²⁺ Contribution

Action potentials do not only occur in neurons: they can be found in muscle cells, as will be considered in ▶ Chap. 4 when we look at how signals can be passed from nerves cells to muscles. In most neurons, voltage-dependent Ca²⁺ channels are also found, which can contribute significantly to the action potential. This is because they often inactivate more slowly than the corresponding Na⁺ channels, causing a much slower repolarization with a plateau phase caused by the Ca²⁺ channels and the increase in intracellular Ca²⁺. This increase can be important in its own right: in ▶ Chap. 4 we will see that this is the trigger for the release of neurotransmitters, which are important in cell-cell communication.

Intracellular Ca²⁺ can also activate other kinds of ion channels (often K⁺ channels that are activated by Ca²⁺), which can lead to a hyperpolarizing undershoot after repolarization due to the increase in p_K. Since the increase in Ca²⁺ concentration lasts much longer than the normal timing of the action potential, the resulting hyperpolarization is some hundreds of times longer than it. This effect is particularly important for heart muscle cells, as we will see in ▶ Chap. 7. This is dependent on having enough Ca²⁺ influx during the action potential and enough Ca²⁺ activated K⁺ channels.

In some case the levels of Ca²⁺ may build up gradually over many action potentials before it is sufficient to activate the K⁺ channels and to cause hyperpolarization (normally called the after-hyperpolarization to distinguish it from the undershoot). This can be a way of obtaining rhythmic bursts of activity punctuated by periods of silent behaviour, particularly in neurons that control rhythmic events, which we will meet in ▶ Chap. 7.

Problems with Ion Channels: Channelopathies

Channelopathies are diseases caused by disturbance to the function of particular ion channels or the proteins that regulate them. In this chapter we have met only a few of the ion channels that are normally present in cells, but we have seen that they play a crucial role in normal healthy physiology. Various conditions are associated with disruptions to the voltage-gated channels we have met in this chapter and these tend to lead to severe symptoms such as temporary paralysis, or seizures, associated with the effect of disruption on the action potential in its role as the fundamental cellular unit of signalling.

Channelopathies have also been associated with some heart conditions. As we will explore further in ▶ Chap. 7, the action potential is essential in controlling the contraction of the heart. Impaired behaviour of the channels can have implications for the timing of action potentials and thus heart beats, a recognised example being the 'long QT syndrome' which can cause an irregular heartbeat especially during exertion. Here, 'QT' refers to timings of the heartbeat as measured using an electrocardiogram (see ▶ Chap. 7), something that is directly related to the electrical activity of the action potential in heart muscle tissue.

3.3 Hodgkin-Huxley Model

The generation and propagation of signals have been studied for at least 100 years: however, the most important piece of work during this time was performed by Hodgkin and Huxley between 1949 and 1952. Hodgkin and Huxley studied the squid giant axon and developed the first quantitative model of the propagation of the electrical signal. This model was such an important step in the understanding of electrical activity that it has been called "the most important model in all of the physiological literature", Keener and Sneyd. Their approach was to model the cell as a simple electrical circuit, ◘ Fig. 3.3. The cell is assumed to have a capacitance C_m that models the insulating effects of a cell membrane that is inherently impermeable to ions. The model includes current inputs through potassium (I_K), sodium (I_{Na}) and other channels (I_L, these last being lumped together and termed leakage). The equilibrium potentials for each ion (or group) are included as voltage sources (E_K, E_{Na}, E_L) and the variable permeability as variable resistors (normally given in terms of their respective conductance: g_K, g_{Na}, g_L). There is also an applied current, which we will examine in more detail later.

Exercise B
Using the simple cell model in ◘ Fig. 3.3 use current balance to derive a differential equation for the electrical properties of the cell.

The Hodgkin-Huxley model starts from applying current balance to the model in ◘ Fig. 3.3, as you did in Exercise B:

$$C_m \frac{dv}{dt} = -\bar{g}_K n^4 (v - v_K) - \bar{g}_{Na} m^3 h (v - v_{Na}) - \bar{g}_L (v - v_L) + I_{app}, \tag{3.2}$$

The potential, v, here is defined as the deviation from the steady state value, measured in units of mV. Notice that in Eq. 3.2 $g_K = \bar{g}_K n^4$ and $g_{Na} = \bar{g}_{Na} m^3$, i.e. the conductances include the variables m, n and h. These refer to the m, n and h gates respectively (actually the gates are named after the variables introduced by Hodgkin-Huxley in their model): when the values are equal to 0, the gates are closed and when the values are equal to 1, they are open (note that the first equation shows that both the m and h gates are needed for Na$^+$ to flow, but only the n gate is required for K$^+$). The powers of m and n in the equation were chosen based on experimental data and are often interpreted as the number of binding sites on the two gates.

The gate parameters have their own governing equations:

$$\frac{dm}{dt} = \alpha_m (1 - m) - \beta_m m, \tag{3.3}$$

$$\frac{dn}{dt} = \alpha_n (1 - n) - \beta_n n, \tag{3.4}$$

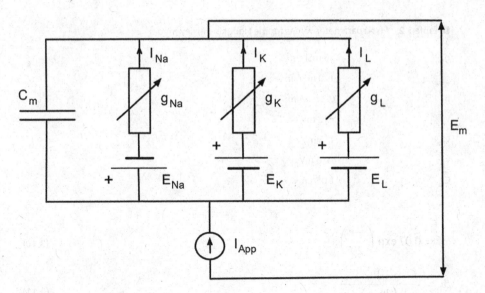

◻ Fig. 3.3 Schematic of Hodgkin-Huxley model

$$\frac{dh}{dt} = \alpha_h(1 - h) - \beta_h h, \tag{3.5}$$

Equations 3.3–3.6 are simply first order kinetic models with forward and backward rate constants like those we met in ▶ Chap. 1. These have a sigmoidal response to a change in potential that matches that seen in experimental data, the powers on the m and n gates result in a steeper slope in their sigmoidal response. The parameters in Eqs. 3.3–3.5 are themselves non-linear functions of the voltage (note that all of these were derived from fitting experimental data):

$$\alpha_m = 0.1\frac{25 - v}{\exp\left(\frac{25-v}{10}\right) - 1}, \tag{3.6}$$

$$\beta_m = 4\exp\left(\frac{-v}{18}\right), \tag{3.7}$$

$$\alpha_n = 0.01\frac{10 - v}{\exp\left(\frac{10-v}{10}\right) - 1}, \tag{3.8}$$

$$\beta_n = 0.125\exp\left(\frac{-v}{80}\right), \tag{3.9}$$

3

▣ Table 3.2	Fixed parameter values for the Hodgkin-Huxley model
g_K	36 mS/cm²
g_{Na}	120 mS/cm²
g_L	0.3 mS/cm²
v_K	−12 mV
v_{Na}	115 mV
v_L	10.6 mV
C_m	1 μF/cm²

$$\alpha_h = 0.07 \exp\left(\frac{-v}{20}\right), \tag{3.10}$$

$$\beta_h = \frac{1}{\exp\left(\frac{30-v}{10}\right) + 1}. \tag{3.11}$$

The remaining parameters all have fixed values as shown in ▣ Table 3.2.

Exercise C
A gate in a cell membrane can have one of two states: closed and open. A simple reaction diagram can thus be written to model changes between the two states:

$$C \underset{\beta}{\overset{\alpha}{\rightleftharpoons}} O$$

where C denotes the probability that the gate is closed and O the probability that it is open.
(a) Write down the two reaction equations governing the behaviour of this gate.
(b) What is the sum of the two probabilities? Hence write down a single equation in terms of the probability that the gate is open. How does this relate to the gate equations in the Hodgkin-Huxley model?
(c) For the equation derived in part (b), write down the steady-state probability of the gate being open and the time constant governing its behaviour.

We will perform some simple analysis of the Hodgkin-Huxley model here to illustrate how a simple model can be used to simulate and to understand the behaviour of a physiological system. The first thing we examine is the steady state behaviour of the variables m, n and h and the time constants. These values are plotted for a range of potential in ▣ Fig. 3.4.

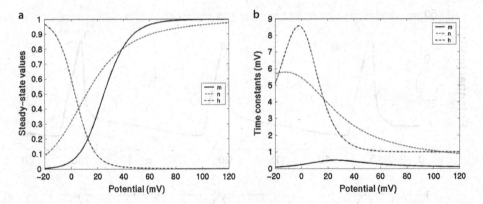

Fig. 3.4 **a** Steady state values and **b** time constants for m, n and h as a function of the difference in membrane potential from its resting value

---

Exercise D

(a) Write down expressions for the steady state probability and time constant for each of the gates in the Hodgkin-Huxley model.

(b) Using the values in ◘ Fig. 3.4, explain how the responses of the different gates give rise to the action potential.

We now apply a stimulus to the system in the form of the applied current: this is equivalent to a neighbouring cell providing a change in potential. If a small stimulus is applied, nothing happens, but if it reaches a threshold value, the action potential suddenly occurs. In ◘ Fig. 3.5, a stimulus is applied at 2 s and left on: the system 'fires' and there is then a refractory period before it fires again. Note that the m gate opens very rapidly, with the n and h gates responding more slowly, as in the model schematic shown earlier. The presence of the applied current in the model does not reflect a physiological current source, rather that the original experiments were done by applying external electrical stimulation to excised nerve cells.

Exercise E

Implement the Hodgkin-Huxely model in a numerical mathematics package using a differential equation solver and reproduce ◘ Fig. 3.5 using an applied current of 5 μA. Determine the smallest current needed to stimulate the model.

It may seem slightly confusing that in ▶ Chap. 2 we considered Na^+/K^+ and Cl^- ions, whereas here we have considered Na^+/K^+ and Ca^{2+} in this chapter. Na^+ and K^+ are the

3

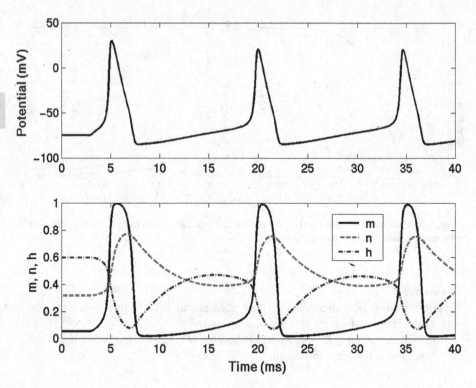

□ Fig. 3.5   Time series of action potential with applied stimulus

dominant ions in the cell's behaviour, whereas the other ions play roles in particular aspects of the cell's behaviour, so have to be considered as and when they are relevant. More detailed models of the cell's behaviour include all the different ions.

Exercise F

Returning to the cardiac myocyte model from Exercise 2F, the following figure shows a schematic of the action potential for this cell.

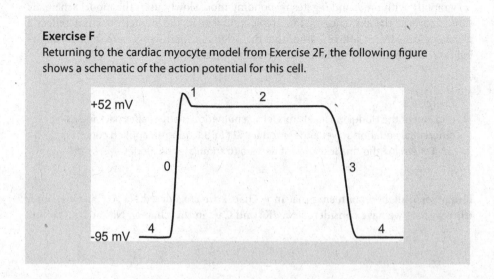

(a) During periods between muscle contraction (phase 4) the membrane is found to a have a high relative permeability to potassium compared to all the other ions. Use this to explain why the membrane potential during this phase is in the range −85 to −95 mV.

(b) The plateau phase (2) is believed to be largely due to the influx of calcium into the cell (through calcium-gated calcium channels). What minimum internal concentration of calcium ions could sustain this plateau? Assume that calcium permeability is far larger than that of all other ions during the plateau.

3.4 Conclusion

In this chapter we have met the action potential, the fundamental electrical signalling mechanism used by the body. You should now be able to understand the cycle of ion movements associated with the generation of an action potential and the role of different ion gates. You should also be able to analyse the system of action potential generation in terms of a simple electrical circuit model.

Cellular Transport and Communication

Contents

© Springer Nature Switzerland AG 2020
M. Chappell and S. Payne, *Physiology for Engineers*, Biosystems & Biorobotics 24,
https://doi.org/10.1007/978-3-030-39705-0_4

In the previous three chapters, we considered how cells were constructed and how they operated. This largely considered cells in isolation, whereas of course cells interact with each other to achieve the tasks that we will consider later. To understand these processes better, we need next to consider how cells are engaged in the transport of substances and how cells communicate with each other.

4.1 Transport

4

One of the important processes that occurs everywhere in the body is transport: getting substances to move from one place to another. The different types of mass transport can be thought of in terms that are analogous to heat transfer, where we have conduction and both forced and natural convection. For example, 'forced' convective transport is achieved by pumping a fluid from one place to another: this is essentially how oxygen is transported around the body, by blood being pumped through blood vessels and we will examine this in ▶ Chaps. 8 and 9. We will consider a number of different processes here very briefly that apply at the cellular level. Although there are only a handful of mechanisms that can be used and the law of conservation of mass always holds, there are many different conditions under which transport occurs.

4.1.1 Passive Transport

We have already considered how concentration differences drive species such as ions in and out of the cell. For a given entity present in solution within a region we can write down, via conservation of mass:

$$\frac{\partial c}{\partial t} = -\nabla \cdot \boldsymbol{J} + f, \tag{4.1}$$

Which essentially says that the rate of change of the concentration c with time equals the rate of production of the substance within the region, f, and the rate at which it leaves across the surface of the region, which is determined by the flux $\boldsymbol{J}$. Intuitively we expect the flux should be related to the concentration gradient, which is the basis of Fick's law:

$$\boldsymbol{J} = -D\nabla c, \tag{4.2}$$

If we substitute that into the conservation law then we arrive at the diffusion equation:

$$\frac{\partial c}{\partial t} = D\nabla^2 c + f, \tag{4.3}$$

where in this case we have assumed that D, the diffusion co-efficient, is constant across space. Since no energy is involved, diffusion is a passive means of transport. The value of D is related to the size and geometry of the chemical species as well as things like temperature and viscosity. If the size of the solute is much greater than that of the solvent then an estimate of D can be found from the Stokes-Einstein equation:

$$D = \frac{kT}{6\pi \mu a}, \tag{4.4}$$

where k is the Boltzmann constant, μ is viscosity and a is the radius of the molecule assuming an approximately spherical shape. We can re-write this in terms of the molecular weight:

$$D = \frac{kT}{3\mu}\left(\frac{\rho}{6\pi^2 M}\right)^{\frac{1}{3}},$$
(4.5)

where, since ρ is approximately constant for large protein molecules we arrive at the result that $D \propto M^{-1/3}$, whereas for smaller molecules, for example respiratory molecules, $D \propto M^{-1/2}$ turns out to be more accurate.

The simplest case of passive transport of a species through a cellular membrane between an external concentration c_o and internal concentration c_i can be described by the formula for flux:

$$J = \frac{DK}{L}(c_i - c_o),$$
(4.6)

where D is the diffusivity of the species *in the membrane*, L is the membrane thickness and K is the partition coefficient. Note that the flux is simply linearly dependent upon the concentration difference. The partition coefficient describes the relative solubility of the species in the membrane as compared to the external (and internal) environments. For a cell this would reflect the typically poor solubility of many solutes in the lipid membrane region compared to the aqueous environments either side of the membrane. It is this relatively poor solubility that means few species pass directly through the cell membrane in any significant quantity. The combination DK/L is often termed the **permeability**, since these terms all relate to properties of the membrane for the species that determine how permeable the membrane appears.

Exercise A

The figure shows a simple 1-dimensional model of a cell membrane through which a species might permeate, with thickness L and diffusivity D. Using the diffusion equation derive an expression for the concentration as a function of x and thus derive an expression for the flux.

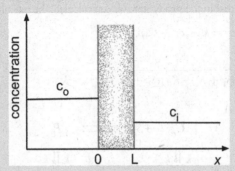

Redraw the figure if the species is twice as soluble in the membrane compared to the external (and internal) solute.

4.1.2 **Carrier-Mediated Transport**

Some substances are insoluble in the cell membrane, hence they cannot readily pass through the cell membrane by diffusion alone, yet can pass through by a process called carrier-mediated transport. This process occurs where a substance combines with a carrier protein at the outer membrane boundary and by means of a conformational change is released on the inside of the cell. This is still essentially a passive transport process since no energy cost is involved.

One of the most important examples of this is the transport of glucose in or out of the cell to provide a source of energy. Although the precise mechanics of this process are not completely understood, a simple model for this transport is that glucose binds with an enzyme carrier protein found in the cell membrane to form a complex. The carrier can change from an 'external' facing state to an 'internal' facing state and vice versa: a conformal change. For example, the glucose outside of the cell can bind with the enzyme carrier protein in its 'external' facing state to form a complex. This complex undergoes a conformal transformation to be 'internal' facing, and then reduces to an 'internal' facing version of the enzyme and glucose now within the cell. The reverse process being possible to remove glucose from within the cell. This is shown in the reaction scheme in ◘ Fig. 4.1, where C is the carrier enzyme and has two states i (internal) or e (external) and can bind with a substrate that is either present inside (i) or outside (e) the cell. When enzyme and substrate bind, they form the complex P, which also has two states. In this scheme binding and reduction occur with rates given by k_+ and k_- respectively, whereas the conformal change occurs at rate k (assumed to be identical in both 'directions'). Notice the similarities with the enzyme reaction models in ► Chap. 1.

Exercise B

Using the reaction scheme in ◘ Fig. 4.1, write down the differential equations for each of the four states of the carrier.

Analysis of this is pretty complicated, but quite possible using the mathematics we used in ► Chap. 1. It turns out that in the steady state we can calculate the rate of supply and demand, J:

$$J = k_- p_i - k_+ s_i c_i = k_+ s_e c_e - k_- p_e, \tag{4.7}$$

$$J = \frac{1}{2} K_d K k_+ C_0 \frac{s_e - s_i}{(s_i + K + K_d)(s_e + K + K_d) - K_d^2}, \tag{4.8}$$

$$
\begin{array}{ccc}
C_e & +S_e & \underset{k_-}{\overset{k_+}{\rightleftarrows}} & P_e \\[4pt]
k \updownarrow k & & & k \updownarrow k \\[4pt]
C_i & +S_i & \underset{k_-}{\overset{k_+}{\rightleftarrows}} & P_i
\end{array}
$$

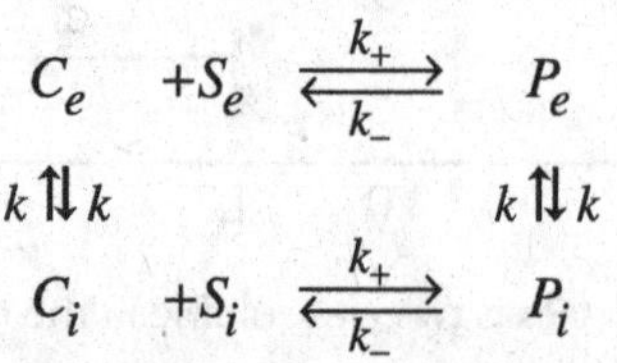

◘ **Fig. 4.1** A model reaction scheme for carrier-mediated transport of a substance S across a cell membrane by means of a carrier C

where $K = k_-/k_+$, $K_d = k/k_+$ and $C_o = p_i + p_e + c_i + c_e$, the total amount of the carrier present in any form. The most important features of this are that:

- at low levels, flux transport is proportional to concentration difference, so it looks like a diffusion process
- there is saturation at high levels of external glucose.

Although the system looks more complicated, the essentials of its behaviour are not that far away from the simple Michaelis-Menten kinetics we met in ▶ Chap. 1.

4.1.3 Active Transport

The processes considered thus far are all passive processes, where ions move down pressure or concentration gradients. There are many processes, however, that require energy to take place, and are thus termed active transport. A case of this is where the concentration of the species inside the cell is actively reduced thus maintaining the concentration gradient required for the passive transport process. For example, glucose is rapidly bound within the cell keeping its concentration low; this phosphorylation of glucose requires hydrolysis of ATP.

Another case is the ion pumps we saw in ▶ Chap. 2, where a Na⁺/K⁺ pump was used to keep the levels of Na⁺ high outside the cell and K⁺ high inside the cell. This requires energy to overcome the gradients and this energy was supplied in the form of ATP. Since the Na⁺/K⁺ pump is essentially a reaction, we can write down a reaction equation for it:

$$ATP + 3Na_i^+ + 2K_e^+ \rightarrow ADP + P_i + 3Na_e^+ + 2K_i^+. \tag{4.9}$$

However, this is a rather simplified version of the actual processes. A more precise schematic is shown in ■ Fig. 4.2. In its dephosphorylated state, Na⁺ binding sites are exposed to the intracellular space; when these ions are bound, the carrier protein is phosphorylated by the release of energy involved in ATP converting to ADP. This exposes the Na⁺ binding sites to the extracellular space, reducing the binding affinity of these sites and releasing the bound Na⁺. A similar process happens, but from extracellular to intracellular, for K⁺.

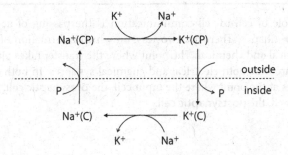

■ **Fig. 4.2** Schematic of Na⁺/K⁺ pump

This can be converted into a mathematical model by writing down three reaction equations (which we won't do here). Note that this is still a simplified version, where the rate of exchange is one Na$^+$ for one K$^+$ ion, rather than the actual three for two. The corresponding differential equations can be written down (again we won't do this here as it is lengthy), where the intracellular Na$^+$ and extracellular K$^+$ are supplied at a constant flux J (and extracellular K$^+$ and intracellular Na$^+$ are removed). In the steady state, this rate can be expressed as:

4

$$J = C_o K_1 K_2 \frac{\left[Na_i^+\right]\left[K_e^+\right]}{\left(\left[K_e^+\right]K_2 + \left[K_i^+\right]K_{-2}\right)K_n + \left[Na_i^+\right]K_1 K_k}, \tag{4.10}$$

where the rate constants are functions of the individual rate constants in the detailed model. The important features of this are that it is very similar to an enzyme reaction, i.e. it has dynamics similar to a Michaelis-Menten reaction, being nearly linear at small concentrations of intracellular Na$^+$ and saturating at large concentrations.

4.2 Cellular Communication

So far, we have considered the cell largely in isolation: we now need to think about how information is passed between cells. Like transport there are processes happening both at the cellular and systemic level. In the latter case the bloodstream is involved as a means to transport signalling chemicals, **hormones**, from one cell or group of cells where the hormone is produced to another region where the signal is 'received'. This is the responsibility of the **endocrine** system and can be seen as a way to broadcast information throughout the body. A similar, but more local system, of communication is achieve by the **paracrine** system, where local 'mediator' chemicals are released by the signalling cell and received by neighbouring target cells.

In this chapter we will be more interested in direct signalling between cells. One way of signalling is **contact-dependent**, where the signalling molecule is physically bound to the cell membrane of the signalling cell. This requires cells to come into contact with the target cell and thus only really applies to mobile cells.

4.3 Synapses

An important role of cell-to-cell communication is the passing of action potentials that we met in ▶ Chap. 3. There are two ways in which information of this form can be passed: electrical and chemical. The point where the transfer takes place is termed a **synapse**: there are thus both electrical and chemical synapses. In both cases there are special structures at the point where the input cell, the **presynaptic cell**, communicates with the output cell, the **postsynaptic cell**.

Exercise C

An increase of extracellular K+ has the same effect as an applied current in the Hodgkin-Huxley model.

(a) Explain why this is the case and derive an expression that relates the necessary change in potassium equilibrium to the size of the applied current.

(b) If a current of 2.3 μA can initiate an action potential, calculate the equivalent change in potassium concentration required.

(c) What relative change in external potassium concentration would give rise to such a change in potassium equilibrium potential? Would a rise in extracellular Na^+ also work?

4.3.1 Electrical Synapses

In Exercise C you have explored a mechanism by which one cell could theoretically initiate an action potential in a neighbouring cell through an increase in potassium in the space between them. A similar mechanism is used at an electrical synapse; the action potential spreads directly to the postsynaptic cell. An electrical synapse is a special region of the cell where postsynaptic cell membrane touches that of the presynaptic cell and the intracellular spaces are connected through special ion channels called gap junctions. The two cells are deliberately put into close chemical contact making it easy for changes in ionic concentration in one to affect another. This is a very rapid means to pass an action potential from one cell to another and is used by cardiac muscle cells, as we will see in ▶ Chap. 6. Apart from being used to conduct electrical signals, it is also the mechanism used to co-ordinate activity in the liver.

4.3.2 Chemical Synapses

At a chemical synapse, an action potential results in the release of a chemical substance (called a **neurotransmitter**). This moves through the extracellular space separating the two cells, and alters the membrane potential of the postsynaptic cell. The best understood chemical synapse is the one between a motor neuron and a muscle cell, termed a **neuromuscular junction**, ▣ Fig. 4.3. The process proceeds as follows:

- The action potential arrives at the presynaptic cell and causes depolarization.
- Depolarization causes Ca^{2+} channels to open.
- Ca^{2+} enters the presynaptic cell.
- Synaptic vesicles fuse with the membrane.
- Neurotransmitters are released into the synaptic cleft.
- Neurotransmitters bind to special channels found on the post-synaptic cell surface.
- Channels open allowing Na^+ and K^+ to cross, permeability of both thus increases.
- Depolarization of post synaptic cell.

4

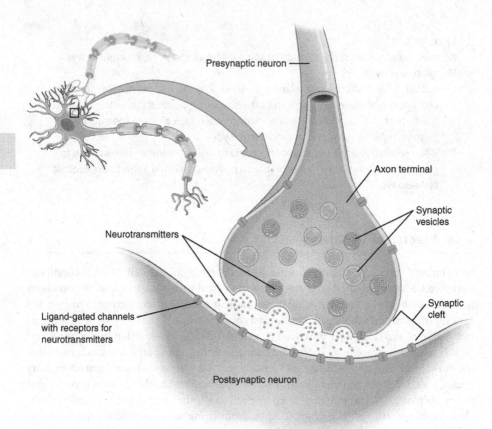

Presynaptic neuron

Axon terminal

Synaptic vesicles

Neurotransmitters

Synaptic cleft

Ligand-gated channels with receptors for neurotransmitters

Postsynaptic neuron

◘ Fig. 4.3 Schematic of neurotransmitter release (this figure is taken, without changes, from Open-Stax College under license: ► http://creativecommons.org/licenses/by/3.0/)

Whilst the absolute permeabilities of both Na⁺ and K⁺ increase by roughly the same amount, the relative permeability to Na⁺ increases more since it started at a lower absolute level. This causes the membrane potential to rise by some 50–60 mV and hence an action potential is generated, the channels closing after a short time (approximately 1 ms).

The neurotransmitter in this case is called **acetylcholine** (ACh). This is stored in units of about 10,000 molecules and so is released in packets, or quanta: a single action potential normally results in the release of more than 100 packets. These packets are stored in large numbers of tiny membrane-bound structures known as synaptic vesicles. ACh is released when the vesicles fuse with the muscle cell membrane at special sites called release sites or active zones, which are only found on the membrane surface opposite the postsynaptic cell. The vesicle membranes are continuously recycled, being filled, emptied and re-filled. Once outside the cell ACh is broken down by an enzyme (choline acetylcholinesterase), allowing the receptor mediated channels to close. The choline produced by the process is then taken back up into the cell and more ACh is synthesized, before being packaged into vesicles. Many other neurotransmitters exist along with a range of different channels that respond to them, allowing for a variety of different chemical synapses with different properties.

We have considered the neuromuscular junction in some detail: chemical synapses between two neurons are essentially the same, although there are some important differences. Probably the most important is that a neuron may receive synaptic connections from thousands of different neurons, unlike a muscle cell that receives an input from only one neuron as we will see in ▶ Chap. 10.

4.4 **Action Potential Propagation**

We have seen how an action potential is generated and how it can be passed from one cell to another, but it is clearly of no use if it remains isolated in one location in the cell. Instead it must be able to propagate through a cell. In particular we need to ensure that an action potential can be transmitted by nerve cells from the brain to the rest of the body (or vice versa), for example along a neuron to the neuro-muscular junction we met earlier. We might imagine that because the action potential is an electrical signal and is associated with the flow of ions it could travel along a cell in the same way that an electrical signal can travel along a wire.

We can build a simple model of this process using the electrical circuit in ◘ Fig. 4.4, which is effectively the model of an electrical cable. In this model we have an infinitely small segment of a cell with action potential propagation in only one-dimension 'along' the cell. We have modelled the interior cell as an electrically resistive medium, with resistance per unit length R_c, and the cell wall as having both capacitance and conductance per unit area of membrane of C_m and $1/R_m$ respectively. Thus we are including the fact that the cell wall is 'leaky' to ions, but we are simplifying the problem by ignoring the contributions from different ions that we had in the Hodgkin-Huxley model. We are also ignoring any resistance to current in the extra cellular space, effectively assuming that there is a larger volume for conductance there than inside the cell. Using this model we can get a general solution for the potential as a function of both time and space (length along the cell in this one-dimensional case), the **cable equation**:

$$\lambda_m^2 \frac{\partial^2 V}{\partial x^2} = \tau_m \frac{\partial V}{\partial t} + V, \tag{4.11}$$

where $\tau_m = R_m C_m$ is the **membrane time constant** and $\lambda_m = \sqrt{\frac{R_m}{R_c}}$ is the **cable space constant**.

4

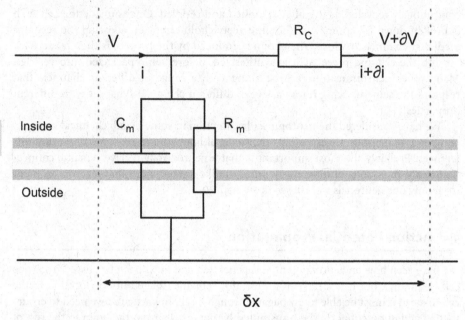

□ **Fig. 4.4** The cable model of passive action potential propagation along a cell

Exercise D

(a) Starting with □ Fig. 4.4 derive the cable equation in Eq. 4.11. Note that by convention R_c is a resistance per unit length of cell, but C_m and R_m are defined per unit area of membrane.

(b) By convention properties of the membrane are quoted as *specific* resistance and *specific* capacitance for a unit area of membrane, r_m and c_m respectively. Properties of the space inside the cell are given as specific resistance of a unit (cross sectional) area of cytoplasm, r_c. For a cell of diameter d, rewrite the membrane time and cable space constants in terms of these properties.

(c) Calculate the membrane time and cable space constants for a cardiac cell with $d = 20\ \mu m$, $r_c = 150\ \Omega$ cm, $r_m = 7 \times 10^3\ \Omega$ cm² and $c_m = 1.2\ \mu F/cm²$.

(d) Consider the behaviour of the system under steady state conditions. If one end of the cell is 100 mV above resting potential at what maximum distance could an action potential be induced, if a threshold (above resting potential) of 20 mV is required to initiate an action potential?

The analysis from Exercise D implies that a change in potential at one point in the cell could initiate another action potential at a distance of the order of 1 mm along the cell. This would be acceptable for the propagation of the action potential in a cardiac muscle cell, which is of the order of 0.1 mm in length, but clearly isn't suitable for nerve cells who have cell bodies metres in length. The analysis in Exercise D was relatively simplistic, and we could be more thorough and specify appropriate boundary conditions and solve Eq. 2.11 analytically. However, this starts to get messy if we want more accurately to model the propagation of an action potential. In fact if we did the analysis more carefully we would not only find that an action potential cannot propagate far along the cell, it also wouldn't do so fast enough for nerve impulses.

A nerve cell cannot act like a wire and simply carry the electrical signal along it. As we have seen, cells are generally very leaky and a lot of the (ionic) current flows out of the cell making them poor conductors. Instead the action potential mechanism itself is employed to achieve a net flow of the signal down the nerve, where an action potential in one region causes another to fire in the neighbouring region of the same cell, called **active propagation**: ◘ Fig. 4.5. An action potential at a point in the nerve cell causes a current of ions that means that the change in membrane potential extends along the cell. This leads to neighbouring regions of the cell reaching a potential above the threshold and thus also initiating an action potential. In this case the action potential could propagate in either direction, in practice an action potential will be triggered at one end and propagate along.

The speed of propagation varies between nerve cells and depends upon how far the effect of the action potential extends along the cell: the larger the region the more rapid propagation occurs. A larger region can be achieved either by reducing the resistance within the cell to the movement of ions, for example by making the cell diameter larger; or by increasing the resistance of the membrane, making it less leaky, so that more of the ionic current flows along the cell. These are both methods to make the cable space constant in Eq. 2.11 larger.

Many vertebrates insulate their nerve cells with a myelin sheath, as is the case for the neuron shown in ◘ Fig. 4.6. The disadvantage of this approach is that the action potential itself relies upon permeability of ions between the inside and outside of the cell. Thus very efficient insulation would actually prevent an action potential from occurring and thus preventing propagation. This is avoided by having breaks in the myelin sheath, called **nodes of Ranvier**. The action potential thus effectively skips from one node to the next.

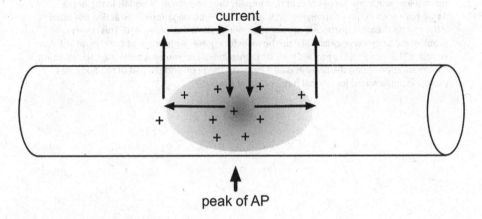

◘ Fig. 4.5 A schematic of active propagation

4

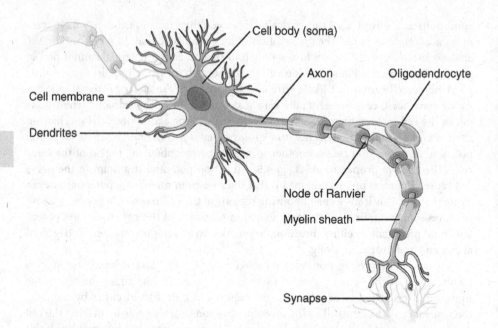

Cell body (soma)

Axon   Oligodendrocyte

Cell membrane

Dendrites

Node of Ranvier

Myelin sheath

Synapse

◻ **Fig. 4.6** Schematic of a nerve cell (neuron): the axon that carries the action potential is insulated with a myelin sheath (this figure is taken, without changes, from OpenStax College under license: ► http://creativecommons.org/licenses/by/3.0/)

Myelin Damage in Multiple Sclerosis

Multiple sclerosis (MS) is a well-known disease of the nervous system that manifests in a range of neurological symptoms such as double vision, muscle weakness and difficulties with sensation or coordination. The name multiple sclerosis refers to the scars (sclerae) that form in the nervous system and are associated with the disease. Although the causes of MS are not well understood, it is associated with a process of 'demyelination': the process of damage to the myelin sheath around nerve cells that impairs the conduction of signals in the nerve. As we have seen in this chapter, myelin is critical to the propagation of the action potential. If the myelin sheath is disrupted, propagation will be slowed or prevented. This is very problematic since nerve cells make up the main functional component of the central nervous system (CNS), as we will see in ► Chap. 10. Disruption of communication in the CNS, affecting communication within the brain and also between the brain and the rest of the body, has profound implications for normal functioning of the body.

4.5 **Conclusions**

In this chapter we have looked at how substances can get in and out of the cell. You should now be able to identify and analyse some of the main mechanisms for transport across the cell membrane. We then looked at the related topic of how cells can communicate with each other. We concentrated on how an action potential can be passed from one cell to another and also how the action potential can propagate along a cell, so that we can achieve rapid signalling over a larger scale.

Pharmacokinetics

Contents

© Springer Nature Switzerland AG 2020
M. Chappell and S. Payne, *Physiology for Engineers*, Biosystems & Biorobotics 24,
https://doi.org/10.1007/978-3-030-39705-0_5

We have already looked at the kinetics of reactions in ▶ Chap. 1, and here we now consider how kinetic modelling can be applied to a whole 'system' such as an organ or the body as a whole. We will see that kinetic models and in particular compartmental models appear in various guises. To introduce the topic, we will look at their use in pharmacokinetics.

5.1 Introduction

Pharmacokinetics, often abbreviated to PK, is concerned with the fate of substances introduced into the body; most obviously this includes therapeutic agents, but might also include things like toxins. Pharmacokinetics is widely used to study substances in the whole body, for example where a drug has been delivered by injection and subsequently blood samples have been taken to determine the plasma concentration as a function of time. From these measurements we are interested in inferring what happened to the drug: how rapidly it was absorbed into tissues, how quickly it was removed from the body etc. However, pharmacokinetics also applies to various imaging modalities where some form of contrast agent is introduced, and we want to calculate a spatially resolved map of absorption etc. We will revisit this when we look at Tracer Kinetics at the end of the chapter. Whilst pharmacokinetics follows what happens to the substance in the body it is also important to know what that substance is doing to the body, which is the aim of pharmacodynamics, but this is something that we will not consider further here.

5.2 ADME Principles

Pharmacokinetics is commonly divided into a number of separate processes, referred to (for obvious reasons) as the ADME scheme, ◘ Fig. 5.1:
- **Absorption** is concerned with how the substance taken up into the blood stream or a specific tissue where it has been introduced.

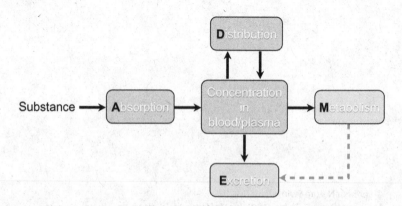

◘ **Fig. 5.1** ADME principles

- **Distribution** is concerned with how the substance is distributed throughout the body, most commonly how it goes from blood to tissue and back again.
- **Metabolism** is the conversion of the substance into other products usually via enzyme reactions.
- **Excretion** is the removal of the substance from the body, for example via the kidneys.

The final two processes together represent **elimination**, since they both result in the loss of the substance. In some cases, we might also need to be concerned with **liberation**, the release of the substance that we are interested in from the formulation that was used to introduce it. Most of these processes, particularly absorption and distribution, will involve some form of transport process to get the substance across membranes in the body. We have already looked at the various passive and active transport processes through which this might be achieved in ▶ Chap. 4.

5.3 Compartmental Models

Whilst there are a number of strategies for pharmacokinetics modelling, we will follow the method of compartmental modelling here. In this approach the substance is considered to be contained by and to move between different compartments. These compartments are very broadly defined and may not represent any specific part of the body, but rather a collection of tissues that share similar properties. The appeal of this method is that the model remains simple and so we have some hope of using it to interpret the data; the disadvantage is that it can be harder then to interpret the model in terms of the underlying physiology.

5.3.1 One Compartment Model

We will start with the simple model shown in ◘ Fig. 5.2; in this case we have a single compartment into which the substance is absorbed and out of which it is eliminated. This is most likely to represent a substance in the blood plasma that is unable to cross into the tissues, thus we will talk about the concentration in this compartment as the plasma concentration. If we assume that a first order output process describes the

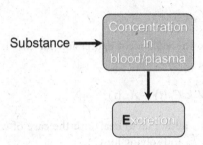

◘ **Fig. 5.2** One compartment model

excretion with a constant k_e then we can write a differential equation for the system, ignoring the input for the time being, as:

$$\frac{dC_p}{dt} = -k_e C_p(t), \tag{5.1}$$

where C_p is the concentration in the compartment. If the substance were introduced by intravenous injection then we can define the input to the system via the initial conditions as the concentration at $t=0$, i.e. $C_p(0)$. Solving this equation gives a simple exponential decay for the plasma concentration:

$$C_p(t) = C_p(0)e^{-k_e t}. \tag{5.2}$$

Thus it would be possible to determine k_e from data of C_p measured through blood sampling. From this we can also define the **half-life** of the substance as the time taken for 50% of the substance to be eliminated: $T_{1/2} = \ln 2/k_e$. We might also be interested in the apparent **volume of distribution** for the compartment, which can be determined simply by dividing the dose given, D, by the initial concentration observed:

$$V_c = \frac{D}{C_p(0)}. \tag{5.3}$$

This is the volume of the compartment into which the substance has distributed itself. We might assume in this case that this volume would equal the blood volume, since we have modelled the drug as only being in the blood. However, the volume of distribution is typically not equal to the blood volume, partially due to the drug only being in the blood plasma, thus not including the substantial volume occupied by blood cells. The volume of distribution allows us to define the **total body clearance**, the rate of elimination per unit of concentration:

$$Cl_{total} = k_e V_c. \tag{5.4}$$

In the case of an intravenous injection it will actually take a short time from the injection for the drug to have distributed throughout the whole circulation. Additionally, excretion will not begin until the substance has reached the right area of the body. Thus $C_p(0)$ will often be taken shortly after injection once a steady 'initial' condition should have been reached.

We can generalise our model for any arbitrary input as:

$$\frac{dC_p}{dt} = -k_e C_p(t) + C_{in}(t). \tag{5.4}$$

It is also useful to write the equivalent expression for the total amount of the substance:

$$\frac{dA_p}{dt} = V_c \frac{dC_p}{dt} = -V_c k_e C_p(t) + A_{in}(t), \tag{5.5}$$

where $A(t) = V_c C(t)$. This allows us to deal with the case of continuous intravascular infusion. Letting $A_{in}(t) = k_0$ and solving for C_p:

$$C_p(t) = \frac{k_0}{V_c k_e}\left(1 - e^{-k_e t}\right).\tag{5.6}$$

From which we can determine the steady state concentration as $t \to \infty$. Once again, we can define $T_{1/2}$: now it tells us how long it takes for 50% of the steady state concentration to be reached, with the steady state being achieved after approximately 5 half-lives.

We can also derive the more general result:

$$C_p(t) = \frac{1}{V_c}\int_0^t A_{in}(\lambda)e^{-k_e(\lambda - t)}d\lambda = \frac{1}{V_c}A_{in}(t) \otimes e^{-k_e t},\tag{5.7}$$

which you might recognise as a convolution as indicated by the $\otimes$ symbol on the right-hand side and thus the need for the dummy variable λ. Our model of instantaneous injection is thus equivalent to $A_{in}(t) = D.\delta(t)$, where $\delta(t)$ is the Dirac delta function. The formulation in Eq. 5.7 allows us to consider more arbitrary intravenous introductions, such as a series of injections or an infusion with a given duration.

Exercise A

(a) Starting with Eq. 5.5 derive the general result in Eq. 5.7.

(b) Show that Eq. 5.6, for the plasma concentration during continuous infusion, can be obtained from this result when $A_{in}(t) = k_0$.

(c) Determine the steady state concentration and consider the roles of k_0 and k_e in reaching that steady state.

5.3.2 Absorption Compartment

Thus far, we have considered only a first-order output. However, the substance may have been administered orally, and so there is likely to have been at least one membrane for it to pass through to get into the plasma, requiring a first order input, as in ◘ Fig. 5.3. Now we have two equations:

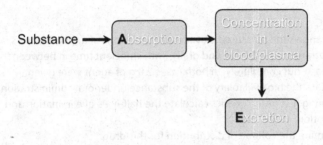

◘ **Fig. 5.3** One compartment plus absorption model

$$\frac{dA_a}{dt} = -k_a A_a(t) + A_{in}(t), \tag{5.8}$$

$$\frac{dA_p}{dt} = k_a A_a(t) - k_e A_p(t), \tag{5.9}$$

where we have introduced a new compartment with concentration A_a. Note that our convolution result above can be applied here again. We could also write the equation for the amount of substance being eliminated:

$$\frac{dA_e}{dt} = k_e A_p(t), \tag{5.10}$$

from which we could solve for $A_e(t)$. This might be useful if we were able to make some direct measurements of elimination, for example urinary sampling.

Exercise B
(a) By modelling A_{in} as a Dirac delta function solve Eqs. 5.8 and 5.9 for plasma concentration C_p in response to a single oral dose.
(b) Sketch the result and comment on its shape.
(c) If $k_a > k_e$ comment on the shape of $\ln(C_p)$ as $t \to \infty$. How might this be used to estimate k_e from plasma-sampled data?

Following Exercise B for a single oral dose of substance $(A_{in}(t) = D.\delta(t))$ the plasma concentration can be written as:

$$C_p(t) = \frac{BD}{V_c} \frac{k_a}{k_a - k_e} \left(e^{-k_e t} - e^{-k_a t} \right), \tag{5.11}$$

where we have also included, B, the **bioavailability** of the substance. This is the fraction of the delivered dose that actually appears in the blood, taking into account losses in the gut through incomplete absorption and metabolism, as well as excretion in the liver. This can be calculated by comparing the area under the plasma concentration-time curve for oral delivery against the same dose of drug delivered intravenously.

Exercise C
The table shows the measured plasma concentrations of a drug that was delivered by both intravenous injection and orally, with sufficient time in between for the first dose to wash-out completely. In both cases 2.0 g of agent were given.
(a) Calculate the bioavailability of the substance under oral administration.
(b) Assuming first order kinetics calculate the half-lives of elimination and absorption.
(c) Determine the volume of distribution for this drug.

IV injection		Oral administration	
Time (h)	Cp (mg/L)	Time (h)	Cp (mg/L)
0.10	190.2	0.17	47.7
0.25	176.5	0.25	61.5
0.5	155.8	0.33	71.5
0.75	137.5	0.5	83.4
1.00	121.3	0.75	87.3
1.50	94.5	1.00	83.5
2.00	73.6	1.50	69.2
3.00	44.6	2.00	54.8
5.00	16.4	4.00	20.3
7.00	6.0	7.00	4.5

5.3.3 Peripheral Compartment

Finally (at least for our purposes) we will consider a substance that also exchanges from the plasma into the tissue: the distribution process. This requires us to add a further compartment to the model, ◘ Fig. 5.4. We still have a 'central' compartment, which includes plasma, but may also include tissues into which the substance rapidly exchanges (such that we are unable to distinguish them from plasma). We have gained a peripheral compartment, to where our drug is distributed, typically an organ or a collection thereof. This combination is generally called a two-compartment model; note that the absorption process is not counted as a compartment even though it behaves like one. The equation for the central compartment is now:

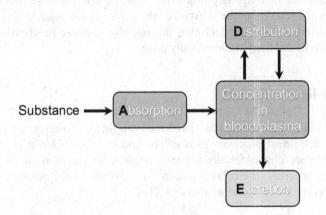

◘ Fig. 5.4   Two compartment model (with absorption)

$$\frac{dA_p}{dt} = k_a A_a(t) - k_e A_p(t) - k_{12} A_p(t) + k_{21} A_g(t), \tag{5.12}$$

and the peripheral compartment:

$$\frac{dA_g}{dt} = k_{12} A_p(t) - k_{21} A_g(t). \tag{5.13}$$

This can be solved reasonably easily for the simple cases of single dose and infusion that we have considered so far to give solutions for C_p that are sums of exponential terms.

Note that in pharmacological studies we generally have access only to the plasma compartment and have to try to determine all the unknowns of the system from that alone, which will be difficult even with this relatively simple system.

5.3.4 Multi Compartment Models

We can build increasingly complex compartmental models if we need to include different peripheral compartments with differing rates of exchange, these will often represent different organs (or groups of organs). The model might also need to account for the metabolism of the substance whilst in the tissue into various products (which may or may not then exchange back into the plasma). Metabolism is simply the result of a reaction, thus the reaction kinetics we saw in ▶ Chap. 1 can be applied and at their simplest match those we have been using here, which should not be a surprise since both are called kinetics. If the metabolic products do end up in the blood stream it might be possible to sample them too and thus to quantify more about the system.

A way to get around (or summarise) the complexity of a multi compartmental model is to return to the convolution expression we had in Eq. 5.7, but now to write the plasma concentration as:

$$C_p(t) = C_{in}(t) \otimes R(t), \tag{5.14}$$

where $R(t)$ is the impulse response of the system (which gives rise to a transfer function in the Laplace or frequency domains). This impulse response function can be interpreted as the fraction of the substance that arrived at any point in time that is still present at some time later. By definition this impulse response function must start at unity at time zero and then monotonically decay.

5.3.5 Non-linear Models

So far all of the compartmental models have been linear. However, we have already met various non-linear processes in reactions and transport. Hence, if we were trying to build more physiologically accurate models, we might need to incorporate some non-linear terms. The most common way to model non-linearities is using the Michaelis-Menten kinetics that we met in ▶ Chap. 1:

$$\frac{dC_p}{dt} = -\frac{V_{max} C_p}{C_p + K_m}. \tag{5.15}$$

This can be regarded as pretty much first order if $C_p \ll K_m$ or zero order (saturated) if $C_p \gg K_m$.

Exercise D

A drug has been developed that is believed to be effective when the plasma concentration is greater than 20 mg/L, but toxic if the plasma concentration exceeds 100 mg/L.

The drug will be administered orally as a single dose D and is assumed to have first order absorption as well as elimination kinetics.

(a) Identify the appropriate equation for the plasma concentration.

A clinical trial is about to begin and an appropriate dosing regime needs to be established. Initial experiments have determined the kinetic parameters for the drug in the table.

(b) Determine the maximum dose that can be delivered.

(c) If the maximum dose is delivered, at what time would the plasma concentration no longer be sufficient for the drug to be effective?

(d) Subsequent study of the drug indicates that the elimination process occurs via an enzyme-mediated process that is more accurately modelled by Michaelis-Menten kinetics with a constant $K_m = 2000$ mg/L. Is the analysis performed in part (c) still valid?

Parameter	Value
F	0.8
V_c	10 L
k_e	0.0028 min^{-1}
k_a	0.2 min^{-1}

The Pharmacokinetics of Painkillers

Paracetamol (acetaminophen) is a commonly used analgesic, i.e. a drug to relieve pain or 'painkiller', for mild pain symptoms with effects typically lasting 2–4 h. After oral administration, it is rapidly absorbed by the gut and has a volume of distribution around 50 L in a typical adult. Paracetamol has a partition coefficient (between water and octanol) of around 3.2, which makes passive diffusion through cell membranes likely for the transport of paracetamol into cells.

After a typical dose (1 g) the peak concentration will reach around 20–30 mg/L (130 µM). When taken four times daily the lowest concentration between doses is around 2 mg/L (13 µM).

The action of paracetamol and how it brings about a painkilling effect is not well understood. Paracetamol is metabolised primarily in the liver, with the products then excreted by the kidneys. The other major class of painkillers are non-steroidal anti-inflammatory drugs (NSAIDs), which include aspirin, ibuprofen and naproxen. As the name suggests (and unlike paracetamol) these have an action to reduce inflammation and are thus often used to treat pain associated with inflammation. Aspirin (acetylsalicylic acid) is a weak acid and is quickly absorbed through the cell membrane in the acidic conditions of the stomach. As much as 80% of a therapeutic dose of aspirin is metabolised in the liver and excreted by the kidneys in a variety of forms. For typical doses the elimination kinetics are first order with half-life between 2 and 4.5 h.

5.4 **Tracer Kinetics**

As we have seen, pharmacokinetics sets out to describe the distribution of agents introduced into the body. Any quantitative measurements typically take the form of concentration samples taken from a specific compartment. With imaging devices, it is possible to get spatially resolved measurements from inside the body, i.e. images of concentration. If we can introduce a **contrast agent**, or **tracer**, that will change the images, we could in principle then measure the distribution of this agent. Two very common examples are **Positron Emission Tomography (PET)** and **Magnetic Resonance Imaging (MRI)**.

In PET, a radioactive tracer is injected into the bloodstream; in principle any biologically compatible substance that can be labelled radioactively labelled can be used. The most common is a sugar called F-18 labelled flurodeoxyglucose (FDG), which behaves similarly to glucose, in that cells take it up. However, it then does not undergo the subsequent (metabolic) reactions that glucose would and instead it gradually accumulates in the organ of interest (before, more slowly, being broken down and removed from the body).

In MRI, the most common tracers are based on gadolinium, a material whose magnetic properties mean that its presence in tissues alters the image acquired using MRI in direct proportion to its concentration. MRI can also exploit blood water as a naturally occurring, or **endogenous**, tracer. This is most commonly applied in the brain where the magnetism of blood water is inverted in the neck prior to imaging in the brain. This process, known as Arterial Spin Labelling (ASL), allows the accumulation of labelled water to be observed in tissue, when it has had time to exchange from the blood into cells.

In all cases it is possible simply to take the image at the right moment, once enough of the contrast agent has accumulated, and hence to visualise the distribution of the agent. Alternatively, it may be possible to acquire time series data of the agent as it is delivered and also removed either by excretion from the cells or through physical decay of the agent signal.

The quantification of delivery from imaging data is based on the principle of tracer kinetics, which is conceptually similar to pharmacokinetics. The contrast agent serves the same purpose as the drug in pharmacokinetics and so we need a description of the input function to the tissue in the imaging region, commonly called the **arterial input function**. This can be obtained by blood sampling, more common in PET, or extracted directly from the images by looking at the time series in larger arteries, as is commonly done in MRI.

Unlike pharmacokinetics, with imaging we have direct measures of the time series of our tracer in the tissue. Rather than getting a single series for a whole organ, we get measurements from each resolution element in the 3-dimensional imaging volume, called **voxels**. Thus, for each voxel we have a time series for the tracer in that volume which will include both blood vessels and tissue. The form of this time series will depend upon what the tracer does when it gets to the tissue: for example, whether or not it all crosses the capillary wall and ends up in the cells or extra cellular space. The simplest case that we might adopt is the one compartment model, making the assumption that the tracer is eliminated according to a first order process once it has arrived in the tissue.

This is often a reasonable simplification for FDG PET, ASL MRI and MRI using a gadolinium contrast agent when imaging outside of the brain. The actual elimination process will be different in each case, representing a combination of metabolism, exchange out of tissue back into the venous circulation and physical decay of the tracer itself. We can write the model for this process as:

$$\frac{dC_t}{dt} = -k_e C_t(t) + F.C_{in}(t), \tag{5.16}$$

where we are measuring the concentration of tracer in the tissue, C_t, we know from an independent measure the AIF, C_{in}, and we have a first order elimination process with rate k_e. Notice that the AIF is scaled by a parameter, F, which quantifies the amount delivered. F has units of blood (carrying the agent) per volume of tissue per unit time; this is in fact a quantify known as perfusion that we will meet again in ▶ Chap. 8. We could solve this equation for the function of C_t with time and use this to quantify perfusion from time series measurements. If we could provide an infusion of tracer, we could simply acquire a single image once steady state has been reached to quantify perfusion. However, for various reasons infusion of the agent generally isn't possible, often because the agent is toxic in too high a dose. For various PET tracers the elimination process is sufficiently slow that it is possible to use a single injection of tracer: the tracer accumulates and remains in the tissue and thus can be imaged sometime (minutes to hours) later for quantification.

Like pharmacokinetics, we can extend the model to more complex cases. For example, we can model what happens if there is a limiting rate at which the tracer leaves the blood into the extravascular space using a two-compartment model. Tracer kinetics theory has a more general result that is applicable in a wide range of more complex tracer uptake situations:

$$C_t(t) = F.C_{in}(t) \otimes R(t). \tag{5.17}$$

This is effectively a further generalisation of the convolution result we met in Eq. 5.14, where again we have an impulse response function $R(t)$, in tracer kinetics this is often called the **residue** function. This function describes the fate of a unit of tracer once it has arrived in the voxel, whether it is washed out in the venous blood, is eliminated through metabolism, decays through some physical process etc. This function has special properties:

- $R(t = 0) = 1$: this simply says that when a unit of tracer arrives it is all there;
- $R(t > 0) \leq 1$: this says that tracer either stays or is removed;
- $R(t_2) \leq R(t_1), t_2 > t_1$: this says that once tracer has been eliminated it cannot come back again.

Note that $R(t)$ only tells us about a unit of tracer: strictly it tells us what happens to a (Dirac) delta of tracer that has arrived in the voxel. Thus Eq. 5.17 models the voxel as a linear time invariant system with the AIF as the input. The observations that tracer accumulates in the voxel is thus a result of the delivery of a tracer, described by the AIF, and that the tracer, once it has arrived, isn't immediately removed from the voxel, as described by the residue function.

$R(t)$ relates to the behaviour of the tracer in the voxel, and thus represents a property (or properties) of the volume of tissue in the voxel. Thus, the shape of $R(t)$ can be interesting in its own right in detecting tissue that is pathological, for example when looking for cancerous tissue. Hence, once we have time series data for each voxel, we can try to estimate $R(t)$ either by writing it as a parameterised function and trying to estimate the parameter values from the data, or performing some form of numerical deconvolution.

5.5 Conclusions

5

In this chapter we have met a general method for the treatment of the body as a whole system in which the behaviour of substances can by described mathematically. You should now be able to use compartmental models built around the ADME principles to write down and to solve differential equations for simple pharmacokinetics problems. We have also seen how the same principles can be applied to measuring the delivery of a contrast agent to tissues, as is commonly used in medical imaging.

From Cells to Tissue

Contents

© Springer Nature Switzerland AG 2020
M. Chappell and S. Payne, *Physiology for Engineers*, Biosystems & Biorobotics 24,
https://doi.org/10.1007/978-3-030-39705-0_6

So far, we have essentially only considered the behaviour of individual cells in the human body. This behaviour is of course very important, but it is not perhaps immediately helpful when we want to study the human body at a larger scale. We will refer to this larger scale as the 'tissue level' behaviour, as a complement to the 'cell level' behaviour. In this chapter, we will consider how we can link these two types of model together. This introduces the very important idea of multi-scale models.

6.1 Introduction

The model of the cell that we have been examining in ▶ Chaps. 1 and 2 of course provides a very neat link to models of tissue. At its simplest level, we can simply consider tissue to be a collection of cells, each of which behaves according to the rules that we set out in ▶ Chap. 2 (conservation of charge, equilibrium potential etc.). It turns out that we can use these rules to understand how tissue behaves at a larger scale. This then provides a good introduction to more general models of how tissue behaves as a material with properties such as Young's modulus.

Note that we will concentrate on soft tissue here, rather than bones and muscle: the main reason for this is that the deformations are larger and there is more interaction between fluids and solids in soft tissue. Exactly the same principles can be applied to bones and muscle, but we will leave discussion of these to the many sources of information that can be found elsewhere. We will start by looking at models of tissue and working through a very simple example, before considering how these can link back to the cell models that we considered earlier.

6.2 Stress-Strain Relationships

Just like any material, human tissue must satisfy the conditions of equilibrium (where forces must balance) and compatibility (where the material must occupy all the available space). These two conditions can be written down in any suitable co-ordinate system, dependent upon the particular problem: we will restrict ourselves to Cartesian co-ordinates here. We will be using exactly the same equations as for any material, but we will see that tissue behaves in quite different ways to traditional engineering materials.

The equilibrium equations in three dimensions are:

$$\frac{\partial \sigma_x}{\partial x} + \frac{\partial \tau_{xy}}{\partial y} + \frac{\partial \tau_{xz}}{\partial z} + F_x = 0 \tag{6.1}$$

$$\frac{\partial \sigma_y}{\partial y} + \frac{\partial \tau_{yx}}{\partial x} + \frac{\partial \tau_{yz}}{\partial z} + F_y = 0 \tag{6.2}$$

$$\frac{\partial \sigma_z}{\partial z} + \frac{\partial \tau_{zx}}{\partial x} + \frac{\partial \tau_{zy}}{\partial y} + F_z = 0 \tag{6.3}$$

where the material has a body force per unit volume in each direction, F_x, F_y, F_z. We are using σ to denote direct stresses and τ to denote shear stresses and the subscripts on the stresses indicate the direction of the force.

The strains are then related to the displacements as follows:

$$\varepsilon_x = \frac{\partial u}{\partial x} \tag{6.4}$$

$$\varepsilon_y = \frac{\partial v}{\partial y} \tag{6.5}$$

$$\varepsilon_z = \frac{\partial w}{\partial z} \tag{6.6}$$

$$\gamma_{xy} = \frac{\partial v}{\partial x} + \frac{\partial u}{\partial y} \tag{6.7}$$

$$\gamma_{yz} = \frac{\partial w}{\partial y} + \frac{\partial v}{\partial z} \tag{6.8}$$

$$\gamma_{zx} = \frac{\partial u}{\partial z} + \frac{\partial w}{\partial x} \tag{6.9}$$

where we are using ε to denote direct strains and γ to denote shear strains. The displacements are u, v and w in the x, y and z directions.

To solve these sets of equations, i.e. to relate forces to displacement, we need to relate stress and strain through the material properties of the tissue: for example using Young's modulus (the ratio of stress to strain), E, and Poisson's ratio (the ratio of strain perpendicular to applied force to strain in the direction of the applied force), v. There are other material properties such as shear modulus, G, and bulk modulus, K, but these are all inter-related. These material properties assume that the material is **linear** and **isotropic** (i.e. the material properties are the same in every direction), which is far from the case for many tissues. We will think about this in more detail later, but let's start by looking at a linear material.

6.2.1 Linear Material

If the material can be assumed to be linear and isotropic, then the equations are just the same as those for any standard engineering material (such as steel). In this case strains in each direction are the sum of the direct and indirect strains:

$$\varepsilon_x = \frac{1}{E}\left(\sigma_x - v\left(\sigma_y + \sigma_z\right)\right) \tag{6.10}$$

$$\varepsilon_y = \frac{1}{E}\left(\sigma_y - v(\sigma_z + \sigma_x)\right) \tag{6.11}$$

$$\varepsilon_z = \frac{1}{E}\left(\sigma_z - \nu\left(\sigma_x + \sigma_y\right)\right) \tag{6.12}$$

Given suitable initial and boundary conditions, any stress and strain field can be solved from this set of equations. A very simple example is given in Exercise A to solve a one-dimensional displacement problem for a soft tissue.

6

Exercise A

Consider a linear tissue where gravity is the only body force and it acts in the negative x direction.

a. Assume: the problem is one-dimensional in the x direction; the material is of height L; the boundary conditions are zero displacement at $x=0$ and zero stress at $x=L$. Show that the displacement of the tissue as a function of x is:

$$u = \frac{\rho g x}{2E}(x - 2L) \tag{A.1}$$

b. If the tissue has density the same as water and a value of Young's modulus of 500 Pa, calculate the displacement of the top of the tissue if it is 1 cm tall (for convenience take $g = 10 \text{ m/s}^2$).

You will find that the displacement is surprisingly large (approximately 1 mm for a tissue of height 1 cm). This is because the value of Young's modulus is very small compared to the kind of values that you may be used to for more common engineering materials (for example, steel has a Young's modulus of around 200 GPa, which is nearly 9 orders of magnitude higher). Soft tissue is inherently very deformable, as you might expect.

6.2.2 **Non-linear Material**

Of course, most tissues are neither linear nor isotropic: this introduces very considerable additional complexity. As we showed in the previous section, human tissue is also not very stiff, which means that the usual assumption of small displacements is no longer valid. The theory behind tissue movement is thus more complex, because we have to consider two frames of reference.

We won't consider non-isotropic behaviour here, as this is very problem-dependent, but we will look very briefly at non-linear behaviour. A very common equation for modelling soft tissue is that proposed by Fung, based on the idea of **strain energy density**, W. The strain energy density function is just the amount of energy stored in the material per unit volume due to strain.

The advantage of writing down the strain energy density is that it provides a single representation of the stress-strain relationship, from which individual stress and strain terms can be calculated. For a linear isotropic material (such as might be assumed for steel or aluminium, for example) with only direct stresses, this would be:

$$W = \frac{1}{2}E\left(\varepsilon_x^2 + \varepsilon_y^2 + \varepsilon_z^2\right)$$

(6.13)

Individual stress terms are calculated by differentiating Eq. (6.13) with respect to individual strain terms, for example:

$$\sigma_x = \frac{\partial W}{\partial \varepsilon_x} = E\varepsilon_x$$

(6.14)

hence Young's modulus. This is the simplest example and anything else does become very complicated very quickly. However, we can relatively easily extend a linear isotropic material to a non-linear isotropic material by adapting the strain energy density function to give:

$$W = \frac{1}{2}\left[a\left(\lambda_1^2 + \lambda_2^2 + \lambda_3^2 - 3\right) + c\left(e^{b\left(\lambda_1^2 + \lambda_2^2 + \lambda_3^2 - 3\right)} - 1\right)\right]$$

(6.15)

The terms λ_i refer to the principal strains, which are simply the ratios of the new 'lengths' of an element to the original 'lengths' in each direction and thus directly related to strain. The constants a, b and c are set to match experimental data. The Fung expression above then models the fact that at small strains, the Young's modulus of tissue is very small (the first term), but that as the strain increases, the effective Young's modulus increases very rapidly (the second, exponential, term), i.e., at larger strain the tissue appears stiffer, not unlike the non-linear behaviour observed in rubber.

Again, we can use these results to derive relationships for stress and strain, but these are quite complicated in comparison with the linear model, so we won't go any further here. We will, however, explore some examples where tissue interacts with other substances in non-linear ways to show how we can use stress and strain to model soft tissue. This will illustrate how we have to consider tissue mechanics very differently to more traditional engineering mechanics.

6.3 Coupled Cell-Tissue Model

We continue our analysis of tissue by going back to our model of the cell from earlier chapters. Consider a simple cell surrounded by a solution, just as we did in ▶ Chap. 2. The cell has both positive and negative ions inside, with concentrations c^+ and c^- respectively, and the extracellular space has a concentration c^*. The resulting pressure is proportional to the difference in internal and external concentrations:

$$p = RT\left(\left(c^+ + c^-\right) - c^*\right)$$

(6.16)

where R and T are the ideal gas constant and absolute temperature respectively. This equation is just like the ideal gas equation, with concentration replacing density.

In ▶ Chap. 2, we noted that the inside of the cell possesses a certain amount of fixed charge, attached to proteins that cannot move across the cell membrane (unlike the ions, which can move through channels). The intracellular concentrations can be written in terms of the extracellular and fixed charge concentrations:

$$c^+ + c^- = \sqrt{\left(c^{*2} + c^{f2}\right)} \tag{6.17}$$

Of course the fixed concentration, c^f, varies with the size of the cell, since it is the total charge that is fixed, and so concentration varies inversely with cell volume. We can relate this to the expansion or contraction of the tissue:

$$c^f = \left(\frac{\phi_0^w}{\frac{dV}{V} + \phi_0^w}\right) c_0^f \tag{6.18}$$

where ϕ_0^w is the tissue water content in some baseline state. This then gives a scaling for the 'new' volume of the cell as a fraction of the 'original' volume: to calculate this we need to define how the cell volume responds to pressure. This requires a model of the tissue, as we examined earlier (hence the reason that we started with a model of tissue before returning to the cell behaviour). Note that we are starting to move across different scales as we develop the model; this is very common in this type of modelling.

The very simplest model is to assume that displacements are (at least relatively) small and that an applied pressure gives rise to a strain. This is based on the definition of bulk modulus (K):

$$K = -V\frac{dp}{dV} \tag{6.19}$$

The volume change will be:

$$\frac{dV}{V} = (1 + \varepsilon)^3 - 1 \tag{6.20}$$

If the strain is small, then we can approximate this by:

$$\frac{dV}{V} = 3\varepsilon \tag{6.21}$$

Hence, using Eq. 6.19:

$$p = 3K\varepsilon \tag{6.22}$$

The factor of 3 essentially comes from the fact that tissue is three-dimensional and so the volume change is approximately three times the strain in each separate dimension.

In its simplest form, we can combine all of the equations above to give:

$$3K\varepsilon = RT\left(\sqrt{\left(c^{*2} + \left(\frac{\phi_0^w c_0^f}{3\varepsilon + \phi_0^w}\right)^2\right)} - c^*\right) \tag{6.23}$$

This gives a relationship between tissue strain and extracellular concentration (we can calculate the pressure from the strain). Note that it is not a straightforward expression to solve (although it can be written in simpler form, as in Exercise B): it is also strongly non-linear. These are both common features of models derived from physiological systems and we often require numerical methods to solve the equations and models that we derive. This does, however, give many interesting types of behaviour that can help to explain some of the phenomena that we see in physiological systems.

Exercise B

a. Show that Eq. (6.23) can be re-written in the form:

$$c^* = \frac{RT}{2p}\left(\frac{\phi_0^w c_0^f}{\frac{p}{k} + \phi_0^w}\right)^2 - \frac{p}{2RT} \tag{B.1}$$

relating concentration to pressure.

b. By sketching this relationship for concentration as a function of pressure, comment on the form of the relationship. How does the relationship behave as the pressure becomes very small?

6.4 Coupled Fluid-Tissue Model

Having considered a very simple model relating cellular behaviour to tissue behaviour, let's now consider the interaction between tissue and fluid more rigorously. This fluid could be water or blood in very small blood vessels. Hence the approach adopted here could be used to model the capillary bed in tissues that we will meet in more detail again in ▶ Chap. 8. It is not feasible to model individual pores or blood vessels, except on a very small scale, so we have to consider an alternative way of describing this. The most common way of doing this is to use the idea of a **porous medium**. This is a concept that is very familiar in soil mechanics, where it is used to describe the drainage of soils with water content.

The early work on soil consolidation (the fact that loading a soil results in a gradual settlement of that soil when it is saturated with water) was performed by Terzaghi in the 1920s and developed further by Biot in the 1940s. The theory builds upon elastic continua theory with the addition of an expression for increment of water content, θ, as a function of stress and water pressure:

$$\theta = \frac{1}{3H}\left(\sigma_x + \sigma_y + \sigma_z\right) + \frac{p}{R} \tag{6.24}$$

where H and R are physical constants. For a linear material, the stresses are related to strains through Hooke's law:

$$\varepsilon_x = \frac{1}{E}\left(\sigma_x - \nu\sigma_y - \nu\sigma_z\right) + \frac{p}{3H} \tag{6.25}$$

$$\varepsilon_y = \frac{1}{E}\left(\sigma_y - \nu\sigma_x - \nu\sigma_z\right) + \frac{p}{3H} \tag{6.26}$$

$$\varepsilon_z = \frac{1}{E}\left(\sigma_z - \nu\sigma_x - \nu\sigma_y\right) + \frac{p}{3H} \tag{6.27}$$

Hence:

$$\theta = \alpha\varepsilon + \frac{p}{Q} \tag{6.28}$$

where:

$$\varepsilon = \varepsilon_x + \varepsilon_y + \varepsilon_z \tag{6.29}$$

is now defined as the total volumetric strain and:

$$\alpha = \frac{2(1+\nu)G}{3(1-2\nu)H} \tag{6.30}$$

$$\frac{1}{Q} = \frac{1}{R} - \frac{\alpha}{H} \tag{6.31}$$

The water content is thus dependent upon the total volumetric strain and the water pressure in a linear way, with the coefficients dependent upon the material properties of the tissue.

The flow of a fluid through a porous medium is then normally governed by Darcy's law, which is again very commonly found in civil engineering:

$$\mathbf{q} = -\frac{\kappa}{\mu}\nabla p \tag{6.32}$$

In this equation, $\mathbf{q}$ is the vector flow rate per unit area (so it has the dimensions of m/s), and κ and μ are the permeability of the porous medium and the viscosity of the fluid respectively. Essentially this equation just says that flow moves down a pressure gradient, with this being proportional to how easily it can pass through the porous medium and inversely proportional to the fluid viscosity.

To solve this equation, we will consider the fluid to be incompressible, which is in fact a very good approximation since blood is largely composed of water, which is essentially incompressible. In this case, conservation of volume gives:

$$\frac{\partial\theta}{\partial t} = -\nabla.\mathbf{q} \tag{6.33}$$

Equations 6.28, 6.32 and 6.33 can be combined to give:

$$\frac{\kappa}{\mu}\nabla^2 p = \alpha\frac{\partial\varepsilon}{\partial t} + \frac{1}{Q}\frac{\partial p}{\partial t} \tag{6.34}$$

which is the equation governing the fluid; the only thing remaining is the relationship between pressure and strain for the solid matrix. In this context these can be written down as:

$$G\nabla^2\mathbf{u} + \frac{G}{1-2\nu}\nabla.\varepsilon = \alpha\nabla.p \tag{6.35}$$

These come from the equilibrium equations and the statement of Hooke's law above. Solving these equations then just requires a set of initial and boundary conditions and a suitable numerical solver.

Exercise C

a. Consider a porous medium in one dimension. Show that the governing equation for pressure is of the form:

$$\frac{\partial^2 p}{\partial x^2} = \frac{1}{c}\frac{\partial p}{\partial t} \qquad (C.1)$$

and derive an expression for the coefficient, c.

b. Consider a material occupying the positive x line, such that the boundary is at $x = 0$ and the material extends infinitely in the positive x direction. Initially the pressure is zero throughout, but at time $t = 0$, a step change in pressure of magnitude P is applied at the boundary. Calculate the pressure in the material as a function of distance and time.

The result in Exercise C is very similar to the heat equation that governs the flow of heat in a one-dimensional passive material. Solutions to this partial differential equation are very standard and importantly are only dependent upon one parameter, as shown in the exercise. This means that they are well understood and can be applied relatively easily, although of course we are assuming linear and isotropic behaviour, so it does become much more challenging if this is not the case.

6.5 Oxygen Metabolism

Blood flow is primarily aimed at the delivery of oxygen to tissue. Oxygen moves across the vessel wall into the surrounding tissue and then diffuses through the tissue. The complex network of blood vessels, which we will examine in ▶ Chap. 8, ensures that every cell is close to a blood vessel: for example, in the brain no brain cell is more than approximately 25 μm from a blood vessel. Despite this, diffusion is required through brain tissue.

The transport of oxygen can be described by the mass transport equation:

$$\frac{\partial c}{\partial t} + \mathbf{v}.\nabla c = D\nabla^2 c - M \qquad (6.36)$$

where the concentration of oxygen, c, is governed by the velocity field, $\mathbf{v}$, the diffusion coefficient, D, and the metabolic rate of oxygen, M. Note that this is a more general form of Eq. 4.3. We are obviously making a lot of assumptions even in writing down this equation: for example, we assume that there is a single diffusion coefficient that is constant in space (i.e. that the tissue is homogenous) and that we can write down concentration of oxygen at the tissue scale.

The final term, metabolism, is in fact not really a constant and depends on a number of factors, most obviously the concentration of oxygen itself. We saw how this relationship works in ▶ Chap. 1. This relationship is often modelled using a Michaelis-Menten relationship, as discussed in ▶ Chap. 1, since the metabolic rate is mostly constant, but does drop off at low concentrations of oxygen. This again illustrates how these simple relationships derived in ▶ Chap. 1 can be applied to more complex systems.

The mass transport equation can be solved numerically, although this is rarely done in full, i.e. with all terms present. More normally some terms are neglected: in particular in tissue the velocity field is often considered to be negligible. This makes solutions more tractable to obtain. A commonly-used version is the Krogh cylinder model, where we consider an axisymmetric cylinder of tissue surrounding a blood vessel. Although a very simple analysis, it is very commonly used in larger models of the circulation. This particular case is considered in Exercise D.

Exercise D

Consider the steady-state, radially-symmetric version of Eq. 6.36 with a negligible velocity field:

$$0 = D\frac{1}{r}\frac{d}{dr}\left(r\frac{dc}{dr}\right) - M \tag{D.1}$$

within a cylinder of radius R_o around a blood vessel of radius R_i. Solve for the concentration field, assuming a fixed value of c_0 at the inner boundary and zero oxygen flux at the outer boundary.

6.6 Multi-scale Models

In this chapter, we have considered a number of models of tissue following previous chapters in which we looked at the behaviour of cells. These are obviously the same thing, but when you look at textbooks, models do tend to be either at the cell level or at the tissue level, because it is easiest to develop models at one of these scales. Often this is absolutely fine for what we want to do.

However, it is possible to relate the behaviour at a cellular level to that at a tissue level. This typically means that the parameters in the tissue level model, such as the diffusion coefficient of oxygen in Eq. 6.36 above, are calculated based on the properties of the individual cells. This is normally done using multi-scale techniques, which can be quite mathematically complicated.

However, such techniques do give an expression for parameters such as diffusion coefficient that are functions of the diffusion of oxygen through individual cells and their arrangement. This can be thought of as a complicated form of averaging properties at the cell scale to give properties at the tissue scale. This mathematical tool does mean that the two types of model can be linked together, which can be very helpful. In particular, it means that if parameters at the cell scale change (for example in pathological conditions) then we can update the parameters at the tissue level in a consistent manner.

6.7 Conclusions

In this chapter, we have examined ways in which we can model both tissue and its interaction with cells, water and blood. The key idea of the chapter is the fact that there is interaction between the different components of body organs: they cannot be considered in isolation, as for many other engineering problems. This does make the task of understanding their behaviour more complicated and we have only examined a few very simple cases to give you the idea of what can be done in this field. There are very many other examples that you will be able to find elsewhere.

Oedema

There are many diseases that affect the behaviour of tissue in different ways. We will just consider one briefly here, which is **oedema**. This often happens after an ischaemic event (such as a heart attack or a stroke). When cells are starved of nutrients, the transport of water (which we looked at in ▶ Chap. 2) can become imbalanced. The resulting change in water transport results in oedema which can be observed in the form of tissue swelling.

There are two main types of oedema: vasogenic and cytotoxic. In the former, more water enters the extracellular space and the tissue appears to swell due to extracellular factors, whereas in the latter changes in metabolism result in cells swelling, and the tissue appears to swell due to intracellular factors. The two processes can both occur in ischaemic stroke and unfortunately appear to be very similar on medical images (such as Computed Tomography), but the former type of swelling is largely reversible whereas the latter is not. Great care thus has to be taken when treating stroke patients to distinguish between them.

Cardiovascular System I: The Heart

Contents

© Springer Nature Switzerland AG 2020
M. Chappell and S. Payne, *Physiology for Engineers*, Biosystems & Biorobotics 24,
https://doi.org/10.1007/978-3-030-39705-0_7

In the ▶ Chap. 6, we considered general tissue behaviour, building on our understanding of how specific cells behave. We now start to consider specific organ systems and how they are constructed and how they function. We will start with the cardiovascular system in this chapter before considering other systems in ▶ Chaps. 8–10

7.1 Overview of the Cardiovascular System

The cardiovascular system, which is also known as the circulatory system, comprises the human heart, blood vessels and blood. It has five main functions:

1. Distribution of O_2 and other nutrients, for example glucose and amino acids, to all body tissues;
2. Transport of CO_2 and other metabolic waste products from the tissues to the lungs and other execratory organs;
3. Distribution of water, electrolytes and hormones throughout the body;
4. Contribution to the infrastructure of the immune system;
5. Regulation of body temperature.

Its overall aim is to maintain the body within well-defined limits, irrespective of external stimuli: this equilibrium is termed **homeostasis**.

A schematic outline of the system is shown in ◘ Fig. 7.1 with the heart at the middle. Although it looks complicated, it is actually relatively simple: the blood flows in a continuous loop, from the right ventricle to the lungs (where carbon dioxide in the blood is removed and exchanged for oxygen), to the left atrium and left ventricle, out to the body tissues (where oxygen passes into the tissue in exchange for carbon dioxide) and back into the right atrium and ventricle. If you get lost, try tracing a path until you get back to where you started from.

Note that, although the lungs are a single entity, the remainder of the body tissues are very widely spread out. The main body tissues in descending order of blood flow rate are shown in ◘ Table 7.1. The circulation is sometimes divided into pulmonary and systemic components, where the **pulmonary** circulation comprises the lungs and the **systemic** circulation all the body tissues.

7.2 Structure and Operation of the Heart

The most important part of the system is the **heart**, which acts as the pump driving blood round the body. The heart is divided into two sections, left and right: the left side pumps oxygenated blood from the lungs to the tissues, whilst the right side pumps de-oxygenated blood from the tissues back to the lungs. A healthy adult heart pumps approximately 5 L of blood around the body every minute. Over the course of a 70-year life span, a heart will beat several billion times and deliver over 100 million litres of blood.

A schematic of the structure of the heart is also shown in ◘ Fig. 7.1. Although it looks complex, again it is relatively straightforward. The main division is into the left and right sides (confusingly pictures of the heart are normally shown as looking front

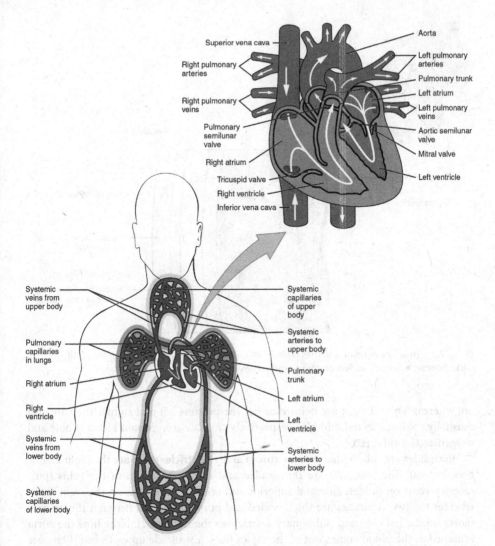

Fig. 7.1   Structure of cardiovascular system and heart (this figure is taken, without changes, from OpenStax College under license: ► http://creativecommons.org/licenses/by/3.0/)

Table 7.1   Distribution of blood between body organs, in descending order

Kidney	Spleen	Skeletal muscle	Brain	Liver	Skin	Bone	Heart muscle	Other
22%	21%	15%	14%	6%	6%	5%	3%	8%

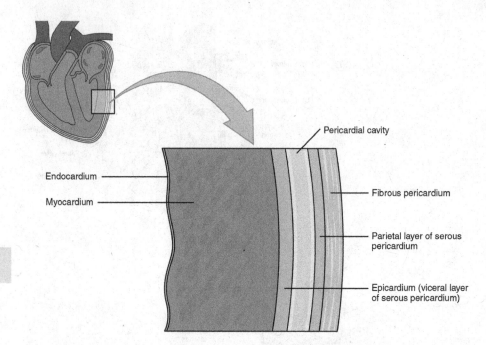

Endocardium

Myocardium

Pericardial cavity

Fibrous pericardium

Parietal layer of serous pericardium

Epicardium (viceral layer of serous pericardium)

7

☐ **Fig. 7.2**   Structure of heart wall (this figure is taken, without changes, from OpenStax College under license: ▶ http://creativecommons.org/licenses/by/3.0/)

on, whereas left and right are defined as per the person's left and right). They are also normally coloured in red and blue respectively since de-oxygenated blood is blue and oxygenated blood is red.

Both sides are subdivided into an **atrium** and a **ventricle**: there are thus four chambers in total. The two atria are thin-walled and receive blood from the veins (pulmonary veins on the left side and superior and inferior vena cava on the right side), whereas the two ventricles are thick-walled and pump blood out through the arteries (aorta on the left side and pulmonary arteries on the right side). Note how the aorta is arched as the blood comes out of the top of the left ventricle upwards but then goes downwards (the small vessels coming off the top of this arch supply the brain and cardiac muscle).

There are also four **valves**: one at the exit from each chamber. The mitral and aortic valves are found on the left side at exit from the atrium and ventricle respectively; the tricuspid and pulmonary valves on the right side similarly. The two valves at exit from the ventricles are semilunar valves with three cusps, i.e. three semicircles that act to cover the cross sectional area, whereas the mitral and tricuspid valves have inextensible fine chords (chordae tendineae) extending from free margins of the cusps to the papillary muscles, which contract during systole, thus acting somewhat like parachutes.

The heart wall is comprised of three layers, as shown in ☐ Fig. 7.2: the epicardium, myocardium and endocardium, from outer to inner. The **endocardium** is a thin layer of cells, similar to the endothelium found in blood vessels; the **myocardium** is the muscle and the **epicardium** is the outer layer of cells. The whole heart is contained within the pericardium, a thin fibrous sheath or sac, which prevents excessive enlargement. The pericardial space contains interstitial fluid as a lubricant.

The cardiac cycle comprises two parts: the resting (or filling) phase, termed **diastole**, and the contractile (or pumping) phase, termed **systole**. During the first phase, blood enters the heart on both sides simultaneously through the veins. The pulmonary and aortic valves are both shut, so the atria and ventricles expand. During the second phase, the heart contracts strongly and the pulmonary and aortic valves open: a pulse of blood thus flows out into the pulmonary arteries and aorta. However, it is important to remember that blood is still flowing in through the veins: the heart must accept this blood and make sure that blood is not forced back into the veins at any point in the cycle.

This is when the separate chambers become important: the tricuspid and mitral valves are shut automatically by the pressure that builds up in the ventricles and so the atria continue to receive blood whilst it is being forced out of the ventricles. The structure of the tricuspid and mitral valves is the reason for this unidirectional behaviour, since they are attached to the ventricular walls and thus permit flow through in one direction only, as shown in ◘ Fig. 7.1. Similarly, when the ventricles relax, the pulmonary and aortic valves act to prevent blood from flowing back into the ventricles: these valves are thus also unidirectional.

Note that the heart valves behave in a very similar way to electrical diodes, where the current can only flow one way. In fact there is always some leakage in the heart valve, just as there is in a diode, but as long as this doesn't get too large, it can safely be ignored. The structure of the heart thus enables it to receive a constant flow of blood through the veins, whilst pumping out blood at regular intervals through the arteries. It is important to remember that the body receives a pulsatile, rather than a steady, flow of blood, although this pulsatile flow is smoothed out rapidly as the blood passes into the vasculature.

The myocardium is the muscle in the heart wall whereby the heart contracts and obviously it must be supplied with blood to function properly. The cardiac muscle has a network of blood vessels to supply it, via the left and right coronary arteries, which branch off the ascending aorta. After passing through the myocardium these vessels feed back to the right atrium via the coronary sinus. The operation of the coronary system is thus vital in maintaining continuous blood supply to the heart muscle.

7.3 Measurement of Cardiac Output

Given the key role of the heart in the operation of the human body, it is obviously important to be able to monitor its behaviour. We will look at a number of ways, but the first one that we will examine is termed Cardiac Output (CO): this is simply the rate at which the heart pumps blood out into the pulmonary and systemic circulations. There are a variety of ways to measure CO, both directly and indirectly.

The first technique is based on **Fick's principle**, which simply equates the absorption of oxygen in the capillaries to the oxygen inhaled in the lungs (effectively balancing the input and output of oxygen in the body). The absorption of oxygen in the capillaries can be calculated as the difference between the oxygen contents in the arteries and the veins: these are normally measured in units of ml_O_2/ml_blood. The oxygen consumption of the body, measured in ml_O_2/min, is the product of this difference and the CO, measured in ml_blood/min: we are essentially using oxygen as a tracer. CO can thus be calculated by:

$$CO = \frac{V_{O_2}}{(O_{2\,arterial} - O_{2venous})} \tag{7.1}$$

Indicator dilution techniques are based on the idea that if a known amount of a substance is injected into an unknown volume, the final concentration allows the volume to be calculated. In Hamilton's dye method a quantity of non-toxic dye is injected into a vein: it mixes with the blood as it passes through the heart and lungs. By taking successive arterial blood samples, the mean concentration can be calculated. Cardiac output is then calculated from:

$$CO = \frac{ADI}{MDC \cdot DFP} \tag{7.2}$$

i.e. the ratio of the Amount of Dye Injected (ADI) to the product of the Mean Dye Concentration (MDC) with Duration of First Passage (DFP).

A variant on this is the **thermodilution** technique, which is the most common method to measure CO. A modified Swan-Ganz catheter with a thermistor at its tip and an opening a few centimetres from the tip is inserted from a peripheral vein such that the tip is in the pulmonary artery and the opening is in the right atrium. A small amount of cold saline is injected into the atrium and this mixes with the blood as it passes through the ventricle and into the pulmonary artery, thus cooling the blood. By measuring the blood temperature, the flow rate can be calculated, since the temperature drop is inversely proportional to the blood flow. This is very similar to the indicator dilution technique, but measuring the dilution of temperature rather than concentration.

More modern non-invasive techniques (i.e. those that do not require the insertion of any probes into the body) use imaging methods to estimate real-time changes in the size of the ventricles. This can be used to give the Stroke Volume (SV): the volume of blood pumped out in one cardiac cycle. The CO can then be calculated, using a simultaneous measurement of the heart rate (HR), from:

$$CO = SV \cdot HR \tag{7.3}$$

Imaging methods, like MRI, have the potential for great accuracy and for beat-to-beat measures of CO, but are of course very much more expensive due to the need for very expensive equipment, despite being non-invasive in nature.

Exercise A

Calculate cardiac output if oxygen consumption is calculated to be 200 ml_O_2/min and arterial and venous oxygen concentrations are 0.21 and 0.16 ml_O_2/ml_blood. If the heart rate is 60 beats per minute, calculate the stroke volume. Comment on your answer.

7.4 Electrical Activity of the Heart

Thus far, we have simply considered the mechanical activity of the heart, by looking at it as a pump and thinking in terms of blood flow. We will now look at the electrical activity of the heart, considering what drives it, in particular what makes it expand and contract, before looking at how we can monitor its performance. This builds upon ► Chap. 3, where we looked at the generation of the action potential.

7.4.1 The Action Potential

The action potential of a cardiac muscle cell, as shown in ◻ Fig. 7.3a, is similar to that of the cells that we examined earlier, but with the addition of a sustained plateau due to the influence of Ca^{2+}. Calcium ions play a very important role in cardiac cells.

Depolarization: This is the first phase of cardiac cell firing: this lasts approximately 2 ms. As in the example shown in ► Chap. 3, the action potential is generated by a sudden transient rise in Na^+ permeability with a subsequent increase in potassium permeability (remember the m, n and h gates). The inward current of Na^+ ions through voltage-gated Na^+ channels becomes sufficiently large to overcome the outward current through K^+ channels and so the cell potential increases (becomes less negative): the membrane thus suddenly becomes much more permeable to Na^+ ions due to the channels. This process then activates more Na^+ channels and the process becomes self-perpetuating (essentially there is positive feedback and the system is unstable).

Plateau: The role of calcium now becomes crucially important. When the membrane potential reaches a threshold, voltage-dependent calcium channels open and there is a large influx of calcium ions. There are also some potassium channels that close upon depolarization (the opposite to the potassium channels discussed in ► Chap. 3). These two effects contribute to the plateau, where the membrane potential decays only very slowly over approximately 200 ms, even though the sodium permeability returns virtually to its resting value.

Repolarization: The next stage shows a more rapid drop. This is due to a gradual decline in the calcium permeability and an increase in potassium permeability. The precise mechanisms are still not fully understood, but the decline in calcium permeability may be due to the effect on the calcium channels of the accumulation of calcium in the cells. By the end of this phase, the outward potassium current returns to its dominant position and the membrane potential returns to its resting value before the next action potential begins.

Since the action potential automatically causes a mechanical response, with the tissue contracting and becoming shorter in length, it might be wondered why there is a need for a more complicated action potential. This is because in cardiac muscle, the duration and strength of the contraction are important parameters that will have to be adjusted dependent upon the circumstances: more control is needed over the action potential and this is achieved through the use of calcium ions. This is

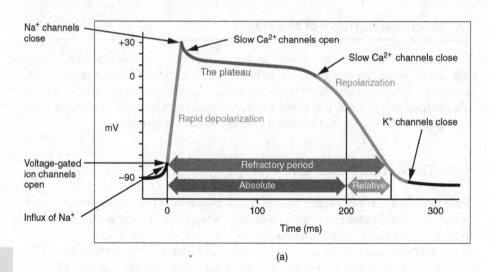

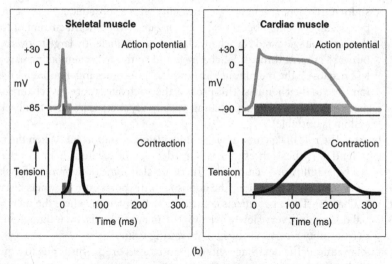

□ **Fig. 7.3**   Action potential in cardiac muscle cells: **a** electrical activity; **b** electrical and tensile behaviour (this figure is taken, without changes, from OpenStax College under license: ▶ http://crea-tivecommons.org/licenses/by/3.0/)

shown in □ Fig. 7.3b, where the prolonged action potential results in a more powerful contraction: calcium ions thus give a lot of **control** over the magnitude of the tensile forces.

7.4.2 Pacemaker Potential

Heart cells in isolation beat rhythmically, i.e. they exhibit a spontaneous action potential that does not require an external stimulus to start the process. They therefore do not have a true resting potential. Like skeletal muscle, the cardiac action potential has a phase where the open potassium channels slowly close (this was the 'undershoot' discussed for the earlier action potential). As the system returns to 'baseline' the threshold is once again reached for depolarization and the whole process restarts. This is referred to as the **pacemaker potential**. In reality the rate of action potentials in cardiac muscle cells varies from cell to cell and so there needs to be some form of co-ordination to ensure that the whole muscle contracts simultaneously.

7.4.3 Cardiac Cycle

The stimulus for the cardiac action potential is provided by the **SinoAtrial Node** (SAN): the cells found in the SAN are often termed **pacemaker cells**. It has an unstable resting potential, as shown in ◘ Fig. 7.4, which decays from approximately -60 mV to a threshold value of -40 mV, at which an action potential is initiated. The rate of decay of this resting potential thus determines the rate of firing and hence the heart rate. In a healthy human at rest, this will result in action potentials being started approximately 70–80 times each minute.

As the SAN cells depolarise, they stimulate the adjacent atrial cells, causing them to depolarise similarly. The depolarisation wave spreads over the atria in an outward-travelling wave from the point of origin, the action potential passing from cell to cell via electrical synapses, as described in ▶ Chap. 4. Both atria contract nearly simultaneously, as do both ventricles shortly afterwards. Co-ordination between atria and ventricles is achieved by specialised cardiac muscle cells that make up the conduction system, as shown in ◘ Fig. 7.5.

The wave is prevented from spreading past the limits of the atria by a fibrous barrier of non-excitable cells: the only excitable tissue that crosses this barrier is the **Bundle of His**. At the origin of this bundle is a mass of specialised tissue about 2 cm long and 1 cm wide called the **AtrioVentricular Node** (AVN). The conduction velocity

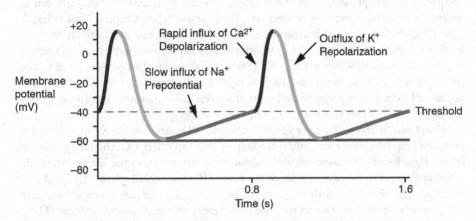

◘ **Fig. 7.4** Action potential at the SAN (this figure is taken, without changes, from OpenStax College under license: ▶ http://creativecommons.org/licenses/by/3.0/)

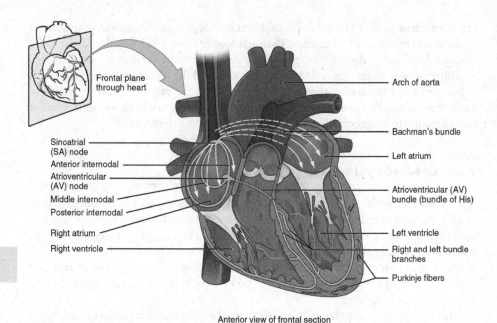

Anterior view of frontal section

☐ **Fig. 7.5** Conduction system in the heart (this figure is taken, without changes, from OpenStax College under license: ▶ http://creativecommons.org/licenses/by/3.0/)

through the AVN is approximately 0.1 m/s, some 10% of that of the atrial cells. The delay that this causes is vital in preventing the ventricles from contracting before the atria have completed their contraction. The impulse from the AVN travels through the Bundle of His, which splits into the left and right bundle branches before dividing into the multiple fibres of the **Purkinje system**. This distributes the impulse over the inner walls of the ventricles causing contraction.

Although the SAN alone would produce a constant rhythmic heart rate, there are regulating factors present, as the heart rate may need to increase, for example during increased physical activity, or to decrease, for example when sleeping. This is largely controlled by the AVN, with most of the changes in heart rate mediated through the cardiac centre in the brain via the **sympathetic** and **parasympathetic** nervous systems, which will be examined in ▶ Chap. 10. Obviously, there are a large number of factors involved in setting the heart rate, such as body temperature, ion concentrations, oxygenation levels, blood pressure and even emotions (although how emotions, such as stress, drive your heart rate is a very complicated subject).

Heart rate increases to raise Cardiac Output (CO) since stroke volume is largely constant (Eq. 7.3). There are only three factors that can affect CO: the filling pressure of the right heart (the **preload**), the resistance to outflow from the left ventricle (the **afterload**) and the functional state of the heart. The last includes heart rate and contractility: the ability of cardiac muscle to generate force for any given fibre length. The stroke volume is known to vary with the ventricular end-diastole volume (EDV), according to the Frank-Starling law of the heart, such that an increase in this volume causes an increase in stroke volume.

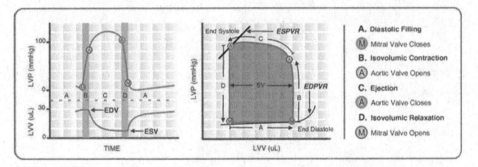

Fig. 7.6 Cardiac pressure volume cycle (LV = Left Ventricle) (this figure is taken, without changes, from Wikimedia under license: ▶ https://creativecommons.org/licenses/by-sa/2.5/deed.en)

■ Figure 7.6 shows the pressure and volume changes during the cardiac cycle, firstly over time (LHS) and then against each other (RHS), for the left ventricle. The second of these plots is used for any cycle, rather like an internal combustion engine, where the area enclosed by the loop is the work done to power the cycle (in the human body this must come from metabolism). Increases in preload raise stroke volume, but increases in afterload decrease stroke volume, as the loop moves upwards and the left side to the right. Increases in contractility move the end systolic pressure-volume relationship (ESPVR) line upwards, thus increasing stroke volume since line D moves to the left.

7.4.4 Introduction to Electrocardiography

The pumping cycle is controlled by a conduction system to give a series of events in a very well-defined order. The conduction system generates current densities through the membrane activity of the heart muscle cells and this is important in how we measure electrical activity in the heart. The ionic currents flow in the thorax, which contains no other sources or sinks and is thus an almost entirely passive medium.

Currents flowing through resistive loads produce voltages: they are of course very small, but, by placing electrodes at different positions on the human body, potential differences can be recorded. These potential differences form the basis of the **Electro-CardioGram** (ECG). There are of course difficulties involved in attempting to measure very small potential differences (of the order of a few mV) on a living human, but we won't look at those here. We will confine ourselves to understanding the production and interpretation of the signal.

The ECG is based on the idea of an equilateral triangle (known as Einthoven's triangle) with the heart as a current source at the centre. Since the potential difference depends upon both the current magnitude and direction, the ECG is a vector quantity. The three bipolar leads, i.e. measurements made between two points, are known as Lead I (right arm to left arm), Lead II (right arm to left leg) and Lead III (left arm to left leg): hence the idea of a triangle, as shown in ■ Fig. 7.7. There are then a further nine unipolar leads, i.e. measurements made at one point: aVR (right arm), aVL

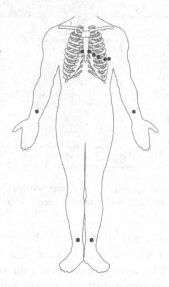

7

◘ Fig. 7.7   Locations of leads used in ECG (this figure is taken, without changes, from OpenStax College under license: ► http://creativecommons.org/licenses/by/3.0/)

(left arm), aVF (left leg) and V1–V6, as shown in ◘ Fig. 7.7, where V1–V6 run from left to right across the chest. Note that this naming system is simply the convention that is used for historical reasons.

Although each of these 12 leads will give a different output, the characteristics of the heart cycle are most clearly shown in lead II, since this pair aligns most closely with the major axis of potential differences in the heart. A typical waveform is shown in ◘ Fig. 7.8. It is labelled at various points: P, Q, R, S and T.

P wave: The depolarisation of the atria prior to atrial contraction causes a small low-voltage deflection, followed by a delay.

QRS complex: The depolarisation of the ventricles prior to ventricular contraction causes a large voltage deflection: this is the largest-amplitude section of the ECG. Although atrial repolarization occurs before ventricular depolarisation, the resulting signal is very small and thus not seen.

T wave: Ventricular repolarization. This last section is much the most variable and often very hard to see.

The sections between these features are close to zero potential, since there are few other sources of potential difference.

A great deal of research has been done to interpret the ECG in terms of providing a clinical diagnosis. The simplest measure that can be extracted is the heart rate, normally taken as the inverse of the time between successive R peaks (the RR interval), measured in beats per minute. A trained clinician will also be able to infer a very large amount of information about the state of the heart from simply looking at an ECG, using changes in the relative timings and amplitudes of different sections of the ECG. Measurement of the ECG is simple and cheap, meaning that it is normally the first

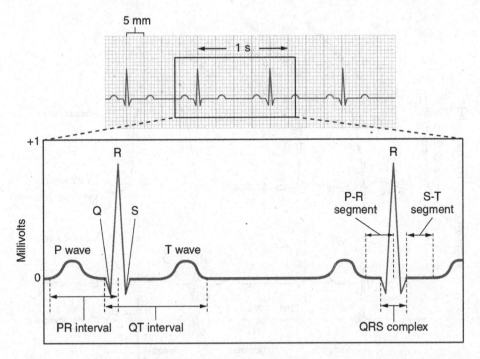

□ Fig. 7.8 Schematic of ECG signal (this figure is taken, without changes, from OpenStax College under license: ▶ http://creativecommons.org/licenses/by/3.0/)

monitoring device attached to a patient. Note that □ Fig. 7.8 shows a perfectly regular heartbeat, which is not what is actually found: in fact, the variability in your heart rate is a good sign and this can also tell us a lot about how the heart is working, as we will see in ▶ Chap. 10.

Exercise B

a. What is the heart rate (in beats per minute) of the patient with the ECG shown in □ Fig. 7.8? Is there any variability from beat to beat?

b. By holding the fingers of one hand against the other wrist, estimate your heart rate in beats per minute. After a short period of light exercise, estimate it again. Comment on why the values are different.

A brief summary of the cardiac cycle is shown in □ Fig. 7.9, relating the ECG to the pressure and volume changes. Check that you can follow the process from both a mechanical and an electrical point of view. There is ejection during systole, which is generated by ventricular depolarisation (shown by the QRS complex), which causes the pressure in the left ventricle to increase rapidly and the volume to drop. When the pressure rises above the aortic pressure, the aortic valve opens and blood leaves the ventricle: the pressure thus drops and when it drops below aortic pressure the

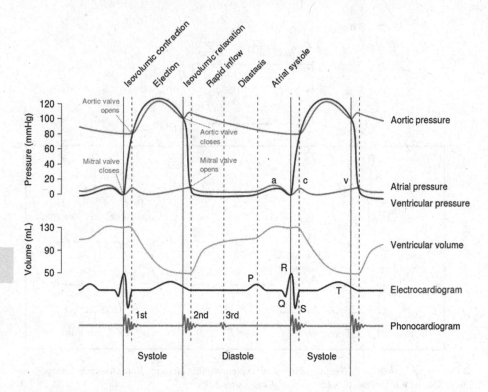

❏ **Fig. 7.9**   Cardiac cycle (this figure is taken, without changes, from Wikimedia under license:
▶ https://creativecommons.org/licenses/by-sa/2.5/deed.en)

aortic valve closes again. Ventricular repolarization occurs during the remainder of systole (shown by the T wave) whilst the ventricles slowly return to their steady state conditions.

Note that there are characteristic heart sounds, which are also shown in ❏ Fig. 7.9. The first heart sound is low frequency and associated with closure of the atrioventricular valves; the second is higher frequency, reflecting a longer ejection period in the right ventricle; and the third is associated with rapid refilling. There is a fourth sound, corresponding to atrial systole, but this is not normally audible.

7.5 **Conclusions**

In this chapter we have looked at the human heart: how it is constructed, how it works as a pump and how this is controlled. We have also looked at how we measure how it is performing, both in terms of the cardiac output and the ECG. In the next chapter we will look at the rest of the cardiovascular system and its interaction with the heart to ensure that the body receives the right amount of blood in the right places.

Heart attacks

Heart disease is caused by an imbalance between the blood flow to the myocardium and the myocardial metabolic demand. This is termed **ischaemic** (meaning a reduction in blood supply). If the coronary arteries block up (normally over a long period through deposits of fatty substances on the wall), their resistance to flow increases, reducing the supply of blood. A **myocardial infarction** (the technical term for a heart attack) means that some of the heart muscle cells die due to this lack of blood supply. This tissue death can cause irregular rhythms, which can be fatal even if there is enough healthy muscle to continue pumping.

After myocardial infarction, the heart can recover, although it will rarely be able to pump as much or as efficiently as before. Nowadays, most patients survive heart attacks, due to rapid treatment, and bypass operations, which open up the blood supply by introducing new pathways to circumvent blocked blood vessels, can reduce the chance of a future heart attack. The heart obviously plays a crucial role in the operation of the human body: if it stops operating for even a very short time, the build-up of waste products and starvation of nutrients leads rapidly to irreversible cell death.

Cardiovascular System II: The Vasculature

Contents

© Springer Nature Switzerland AG 2020
M. Chappell and S. Payne, *Physiology for Engineers*, Biosystems & Biorobotics 24,
https://doi.org/10.1007/978-3-030-39705-0_8

In the previous chapter, we looked at the centre of the cardiovascular system, the heart. The heart is responsible for receiving blood and pumping it out again. This blood passes round the body through the vasculature, which is the complete set of blood vessels in every part of the human body. In this chapter, we will look at these blood vessels, how we can model them mathematically and how they contribute to the maintenance of homeostasis.

8.1 Anatomy of the Vascular System

Blood vessels divide into five main types: arteries, arterioles, capillaries, venules and veins. Blood passes through them in that order and they can be distinguished by differences in size, characteristics and function. Blood exiting the heart first flows through the **aorta**, which divides into **arteries**. These divide in turn into **arterioles** then **capillaries** before joining together in **venules** before entering the heart from the **veins** via the **venae cavae**. A schematic is shown in ◘ Fig. 8.1, where we will explain many of the details shown in later sections. The vessels shown in bright red are oxygenated, whereas the vessels shown in dark red are de-oxygenated, since the oxygen carried by the blood has been transferred to the surrounding tissues in the capillaries.

The **arteries** are thick-walled vessels that expand to accept and temporarily to store some of the blood ejected by the heart during systole and then to pass it downstream by passive recoil during diastole. As the arteries branch off, their diameter decreases,

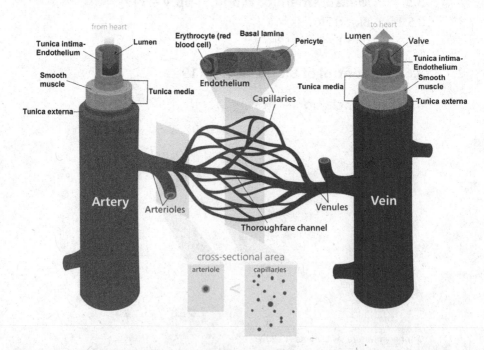

◘ **Fig. 8.1**   Schematic of different blood vessel types and structure (this figure is taken, without changes, from OpenStax College under license: ► http://creativecommons.org/licenses/by/3.0/)

but in such a manner that the total cross-sectional area increases several-fold. The arteries have a relatively low and constant resistance to the flow.

The artery walls comprise three layers: the tunica intima (or tunica interna), the tunica media and the tunica adventitia (or tunica externa), which are simply the inner, middle and outer layers. The middle layer is mainly smooth muscle and is usually the thickest layer: it is important as it both supports the vessel and changes diameter to regulate the blood flow and blood pressure. A schematic of the artery and vein wall structures is also shown in ◘ Fig. 8.1.

The **arterioles** have much thicker walls in proportion to their size than the arteries and have a much greater flexibility to change their diameter and hence their resistance to flow, due to them having lots of smooth muscle. The overall cross-sectional area is again much larger than the arteries but they have a high and changeable resistance, which allows for regulation of blood flow.

The **capillaries** are the smallest vessels with very thin walls that contain no smooth muscle: their resistance is thus largely unchanging. The cross-sectional area is now at its largest: the flow velocity is very small and this is where most of the exchange processes occur. The density and distribution of the capillaries depends upon the requirements of the local body tissues: where tissue consumes lots of oxygen, there will be a greater density than where tissue consumes less. This consumption level is termed the metabolic rate: tissues such as skeletal muscle, liver and kidney have a high rate and thus very many capillaries.

A reverse branching process occurs as the flow enters the **venules** and the **veins**: the vessels branch back together again and the flow velocity increases as the cross-sectional area decreases. The venous vessels have very thin walls in proportion to their diameters: as a result they are very compliant and increase in volume significantly as the pressure changes (unlike the arterial and capillary vessels). They normally contain approximately 70% of the total blood volume at any one time. Just as the arterioles are used to control the resistance, so the venous vessels are used to control the blood volume.

There are valves in medium and large veins to prevent blood from flowing backwards, since the veins are mostly travelling up towards the heart and there is very little pressure to force the blood forwards. Venous return primarily depends upon skeletal muscle action, respiratory movements and the constriction of smooth muscle in the venous walls.

◘ Table 8.1 summarises the structural properties of the different parts of the vascular system in more detail: we will consider how these affect the behaviour of the

◘ **Table 8.1** Structural properties of the vascular system

	Aorta	Arteries	Arterioles	Capillaries	Venules	Veins	Venae cavae
Internal diameter	25 mm	4 mm	40 μm	7 μm	40 μm	7 mm	30 mm
Wall thickness	2 mm	1 mm	20 μm	1 μm	7 μm	0.5 mm	1.5 mm
Length	0.4 m	20 mm	2 mm	0.5 mm	0.5 mm	25 mm	0.3 m
Number	1	280	20×10^6	16×10^9	160×10^6	260	1

cardiovascular system once we have considered the mechanics of blood flow through vessels in more detail. Note that the number of branches is very large, as the cardiovascular network must cover the entire body.

8.2 Blood

Blood consists of two elements: plasma and blood cells. Approximately 40–45% of the volume of blood is occupied by blood cells that are suspended in a watery fluid called plasma. The fraction of blood volume occupied by cells is called the **haematocrit**. The plasma consists of ions in solution and many plasma proteins. By weight it comprises about 90% water and 7% plasma protein, with the remainder made up of other organic and inorganic substances. Blood cells are divided into **erythrocytes** (red blood cells), **leucocytes** (white blood cells) and **platelets**, as shown in ◘ Fig. 8.2. Although they are all produced in the red bone marrow, they have different functions and white blood cells come in a large number of types.

The red blood cells are by far the most numerous and they contain the haemoglobin that is responsible for the transport of oxygen. They are biconcave discs and deform easily to pass through the capillaries, since they are approximately 7 μm in diameter. White blood cells defend the body against infection whilst platelets play an important role in haemostasis (the formation of blood clots in response to damage to the vessel wall). An adult human contains approximately 5.5 L of blood.

8.3 Haemodynamics

Haemodynamics is the study of the flow of blood. Since blood is a fluid, it can be treated like any other fluid. There have been many models proposed to simulate blood flow and we will examine a couple of these here, as they lead to results that are very

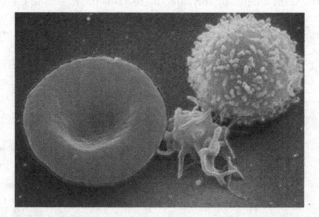

◘ **Fig. 8.2**   Red blood cell, platelet and white blood cell, isolated from a scanning electron micrograph (this figure is taken, without changes, from OpenStax College under license: ▶ http://creativecommons.org/licenses/by/3.0/)

commonly applied. The first type of model is a compartmental model, which is a very similar idea to the type of model that we examined in ▶ Chap. 5. We will then examine the flow in an individual vessel in more detail as this introduces the idea of a pulse wave.

8.3.1 Compartmental Models

We will start our development of a compartmental model by considering a single vessel. We assume that the fluid is Newtonian, that the flow is laminar and steady and that the vessel in which it is flowing is straight and rigid. Although this might seem to be a lot of assumptions to make, it actually gives us a very important result.

In Exercise A, you will derive the following relationship between the pressure difference along a blood vessel and the flow rate through it:

$$\frac{\Delta p}{q} = \frac{8\mu L}{\pi R^4} \tag{8.1}$$

where the viscosity of blood is μ and the vessel has radius R and length L. This is in fact a fairly standard result for any fluid and turns out to be surprisingly useful in haemodynamics. It introduces us to the concept of **hydraulic resistance**, which says that when a pressure difference is applied between two ends of a blood vessel, a flow through the vessel will result that is linearly proportional to this pressure difference. The ratio between applied pressure difference and resulting flow is then termed hydraulic resistance. This is given by the RHS of Eq. 8.1 and is only dependent upon the vessel geometry and the viscosity of the fluid. Blood flowing in large vessels has a viscosity approximately three times that of water.

Exercise A

The equation governing flow in a cylinder is:

$$\frac{\partial (r\tau)}{\partial r} = r\frac{dp}{dx} \tag{A.1}$$

where r is radial position, p the fluid pressure (which is only a function of axial distance, x) and shear stress for a Newtonian fluid is given by:

$$\tau = \mu\frac{\partial u}{\partial r} \tag{A.2}$$

Given that the fluid velocity must be zero at the wall, where $r = R$, and that there must be zero gradient at the origin, show that the velocity profile of the fluid can be given by:

$$u = \frac{1}{4\mu}\left(r^2 - R^2\right)\frac{dp}{dx} \tag{A.3}$$

Hence calculate the flow rate and show the result given in Eq. 8.1.

This idea of resistance to flow, which is simply a measure of the friction caused by viscosity in the flow, is very similar to electrical resistance. In a resistor, when a potential difference is applied between the two ends, current flows, linearly proportional to this potential difference.

We can thus use this idea to develop something called an **equivalent electrical circuit**. In the same way that we can write the equation for a resistor (i.e. Ohm's law):

$$\frac{\Delta V}{I} = \mathcal{R} \tag{8.2}$$

we can also write:

$$\frac{\Delta p}{q} = \mathcal{R} \tag{8.3}$$

where the equation describing electrical current flow is mathematically identical to that describing blood flow in a blood vessel.

We can thus draw an electrical equivalent circuit for any number of blood vessels in exactly the same way that we draw an electrical circuit for any number of resistors. In this analogy, voltage (strictly potential difference) equates to pressure difference, current relates to blood flow and the ratio of the two is resistance (electrical resistance or hydraulic resistance). The fact that there is such a close correspondence between the two is why the analogy is used so frequently in this and other contexts.

Hydraulic resistances can be put in series and in parallel in exactly the same way as electrical resistances (see Exercise B). The arteries, arterioles, capillaries, venules and veins flowing in each body organ are in series, since blood flows through them consecutively, whereas the body organs are predominantly in parallel, since blood flows through them simultaneously.

> **Exercise B**
> Show that a network of N blood vessels, each of resistance $\mathcal{R}$, has an overall resistance $\mathcal{R}/N$ when placed in parallel.

Exercise D asks you to calculate the resistances of all of the different parts of the vascular network. You will find that the resistance is dominated by the arterioles: this is why they are predominantly used to adjust the overall vascular resistance. As their resistance drops, flow rate increases for fixed blood pressure or blood pressure drops for fixed blood flow. They do, of course, only adjust the resistance within certain limits, since blood vessels can only change resistance through changes in diameter. The blood flow through each body organ can also be altered by adjusting the relevant arteriolar resistance: since the organs are in parallel, the blood flow can be changed for one organ without having a significant effect on the remainder of the vascular system (and of course an increase can also be provided by an increase in cardiac output).

We have made a large number of assumptions to derive Eq. 8.1, none of which are strictly true. The most important are that the flow is not steady (remember that the heart is a pulsatile pump) and that the vessel radius is not constant. The effects of pulsatility and an expanding wall diameter are often modelled by means of two extra electrical components.

Inductance is used to model the inertia of the fluid whereas **capacitance** is used to model the storage of blood in vessels, due to the elasticity of the vessel walls: normally called **compliance**. This storage is a very important feature of the vascular system: in fact, the arteries and veins behave more like balloons with a single pressure, rather than resistive pipes with a continuous decrease in pressure.

The **inductance** of a blood vessel is found from solving the Navier-Stokes equations (the equations that govern the behaviour of all fluids) under certain conditions. For simplicity, we will just quote the result here, although it can be derived from the equations later in this chapter:

$$I = \frac{\rho L}{\pi R^2} \tag{8.4}$$

where ρ is the density of blood. Note that we will use I for inductance, rather than L, for obvious reasons.

The **compliance** of a blood vessel is very simply defined as the rate of change of volume with pressure:

$$C = \frac{dV}{dp} \tag{8.5}$$

where we normally define the pressure here as the difference between internal and external pressure (rather than the difference between inlet and outlet pressure). Compliance is analogous to capacitance, which is defined as the rate of change of charge with potential difference.

The compliance depends upon the geometry and properties of the vessel wall. Exercise C asks you to derive the following expression for compliance:

$$C = \frac{3\pi R^3 L}{2Eh} \tag{8.6}$$

This assumes that the vessel wall is of thickness h and made up of a linear elastic material with Young's modulus E and Poisson's ratio of $1/2$. This value of Poisson's ratio relates to an incompressible material, which is in fact a good approximation for the tissue that makes up a vessel wall, as we discussed in ▶ Chap. 6.

Exercise C

a. Show that the compliance of a vessel is given by Eq. (8.6). Assume that the material is linear and elastic with Young's modulus E and Poisson's ratio of $1/2$. The incremental stress components can be taken to be those for a thin-walled vessel:

$$d\sigma_\theta = dp\frac{R}{h} \tag{C.1}$$

$$dσ_z = dp\frac{R}{2h}$$ (C.2)

$$dσ_r = 0$$ (C.3)

b. Show that a network of N capacitors, each of capacitance C, when placed in parallel has total capacitance CN.

We can combine all three of these electrical elements to derive the equivalent circuit shown in ◘ Fig. 8.3 for any type of vessel. Note that we have placed the capacitor in this equivalent electrical circuit at the outlet, but other models will place it in the middle or at the inlet: there is no unique way of doing this. We can use the values given in ◘ Table 8.1 to calculate values of resistance, inductance and compliance for every type of blood vessel, as in Exercise D.

8

Exercise D
Calculate the values of resistance, inductance and compliance for all the different types of vessel listed in ◘ Table 8.1. Assume that vessels have a Young's modulus of 3 kPa and that blood has a density of 1050 kg/m³.

Although we have only considered a single vessel, larger networks of vessels can be built by adding the equivalent circuits in series and parallel as necessary. It is common to simplify the larger networks by neglecting some resistances and capacitances and merging others.

One simple realistic equivalent circuit model of a body organ comprises seven components, as shown in ◘ Fig. 8.4. Obviously this requires some knowledge of the physiology and a number of assumptions. Note that we nearly always neglect capillary inductance and compliance, as done here, since these tend to be very small compared to the other components. The simplest model of the systemic circulation then becomes a series of organs in parallel, where each body organ is supplied by the aorta and drains into the venae cavae.

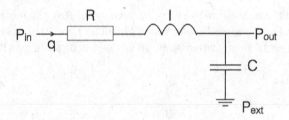

◘ **Fig. 8.3**   Equivalent electrical circuit for blood flow

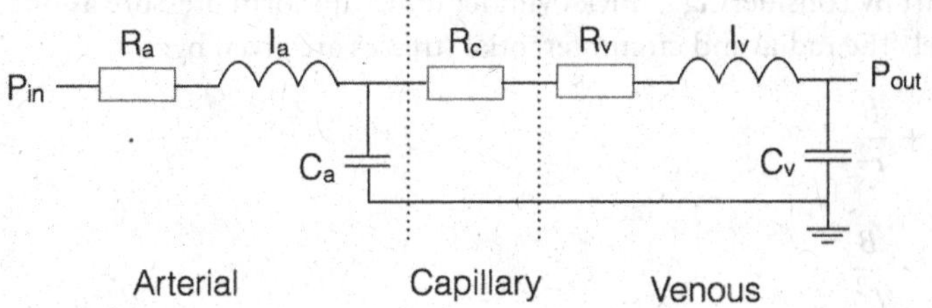

Fig. 8.4 Approximate equivalent electrical circuit for body organ

The compliance of the arteries is vital in converting the pulsatile nature of the flow exiting the heart into a continuous flow. During systole, the flow of blood into the arteries is greater than that exiting into the arterioles, so the arteries expand, contracting during diastole. There is a storing of energy during the first stage, which is then used to propel the blood forward during the second stage. By the time that the flow has reached the capillaries, it is virtually steady state, as you will show in Exercise E.

Exercise E

a. Derive an expression for the flow through the capillaries in terms of the pressure difference between inlet and outlet for the circuit shown in ◘ Fig. 8.4. Use phasor notation where the impedance of an inductor is $Z = i\omega L$ and the impedance of a capacitor is $Z = 1/i\omega C$. Assume for simplicity that venous (outlet) pressure is negligible.

b. Explain how this circuit acts like a low-pass filter. Why is this important?

We can derive more complex equivalent circuits using more detailed models of the fluid flow and the vessel wall. In particular, the vessel wall is not a passive elastic material and the relationship between pressure and volume is more like an exponential rise than a straight line (so blood vessels get less compliant as they get bigger, as you would expect from Fung's model in ▶ Chap. 6). Although this can be modelled using a capacitance that varies with pressure, the model then becomes non-linear and considerably harder to analyse.

8.3.2 Pulse Wave Model

We next consider a single vessel again, but in more detail and relaxing the assumption that the vessel wall must be rigid. Some of the material that we covered in ▶ Chap. 6 will be useful here as we consider the behaviour of the vessel wall. There is here a strong interaction between the fluid and the structure due to the fact that the fluid has a variable pressure. This is very similar to the theory of thin-walled and thick-walled vessels, used widely in standard pressure vessel theory, for example. Note that we will be using $r-\theta-z$ (cylindrical polar) co-ordinates here.

Let's start by considering a thick cylinder under uniform pressure as our model of a blood vessel. The radial and circumferential stresses are given by:

$$\sigma_r = A + \frac{B}{r^2} \tag{8.7}$$

$$\sigma_\theta = A - \frac{B}{r^2} \tag{8.8}$$

where A and B are constants, set by the boundary conditions. Note that these come from solving the equilibrium and compatibility equations in cylindrical polar co-ordinates (these are standard results, so we won't derive them here).

If we assume that we have a blood vessel with blood at a uniform pressure p inside it and p_{ext} outside it, these equations turn into:

$$\sigma_r = \frac{pR^2 - p_{ext}(R+h)^2 + R^2(R+h)^2(p_{ext}-p)/r^2}{(R+h)^2 - R^2} \tag{8.9}$$

$$\sigma_\theta = \frac{pR^2 - p_{ext}(R+h)^2 - R^2(R+h)^2(p_{ext}-p)/r^2}{(R+h)^2 - R^2} \tag{8.10}$$

where the wall has thickness h and inner radius R.

8

Exercise F
Show that the expressions in Eq. 8.9 and 8.10 are compatible with the pressure boundary conditions.

At this point, we need to make an assumption about the stress or strain in the axial direction to calculate the displacements. There are two very common choices: plane stress (where the stress in the axial direction is zero) or plane strain (where the strain in the axial direction is zero). We will consider these briefly in turn.

Plane Stress
In the first case, there is no stress in the axial direction and the vessel is allowed to expand in this direction as it wishes. The circumferential strain in cylindrical polar co-ordinates is given by:

$$\varepsilon_\theta = \frac{u}{r} = \frac{1}{E}(\sigma_\theta - \nu\sigma_r) \tag{8.11}$$

where the material typically has a Poisson's ratio, ν, of 1/2 (the value for an incompressible material). Substituting the expressions for stress into this equation at the inner radius then gives the internal displacement of the vessel wall. The resulting expression is long and complicated, so we won't quote it here; however, the main thing to note is that the displacement is linearly proportional to both the internal and the external pressure. This should be no surprise, since we assumed a linear material;

however, it does still illustrate that even with these simple assumptions the final expression can be highly complicated.

Plane Strain

In the second case, there is no strain in the axial direction and the vessel is constrained to expand only in the radial direction. The circumferential strain in cylindrical polar co-ordinates is thus:

$$\varepsilon_\theta = \frac{u}{r} = \frac{(1 - v^2)}{E}\left(\sigma_\theta - \frac{v}{1 - v}\sigma_r\right) \tag{8.12}$$

Substituting the expressions for stress into this equation again gives a linear relationship between pressures and wall displacement, but with different coefficients (it can be shown that the plane stress condition gives a smaller displacement than that for plane strain for the same applied internal pressure).

Having derived this result, we can simply say that both assumptions yield the result that displacement is proportional to pressure. Since the vessel wall inner radius is not zero at zero pressure, we introduce an offset and turn the expression into the form:

$$p - p_0 = G_0\left(\frac{R - R_o}{R_o}\right) \tag{8.13}$$

where we define the wall stiffness as G_o and the vessel wall has radius R_o at pressure p_o. Notice that we are slightly abusing our notation to write R as the variable inner wall radius now (i.e. equal to the original radius plus the displacement). This allows us to relate the pressure inside the vessel to the wall radius.

The pressure inside the blood vessel will be determined by the flow of blood in the vessel, so, having considered the relationship between wall displacement and applied pressure, we turn to the governing equations for flow inside the vessel.

We start by assuming an axisymmetric straight vessel and write down the equations for **continuity** and **momentum** as:

$$\frac{\partial A}{\partial t} + \frac{\partial (AU)}{\partial x} = 0 \tag{8.14}$$

$$\frac{\partial U}{\partial t} + U\frac{\partial U}{\partial x} = -\frac{1}{\rho}\frac{\partial p}{\partial x} + \frac{f}{\rho A} \tag{8.15}$$

where the vessel has cross-sectional area A, and the fluid has average velocity U, pressure p and density ρ. f denotes the friction term caused by the viscosity of the fluid. The first equation (continuity) balances the storage of fluid due to changes in cross-sectional area with the rate of change of flow along the vessel. The second equation can be thought of as an application of Bernoulli's equation with an added term for the friction of the fluid (this is what dissipates energy in a real fluid).

Note that there are four variables in these two equations: we therefore need two more equations. We can calculate the frictional force if we estimate the velocity profile of the fluid through the vessel: it is normal to guess a polynomial profile, as you will see in Exercise G.

Exercise G

Assume that the velocity of the fluid through a blood vessel is of the form:

$$u(r) = U_{max}\left(1 - \left(\frac{r}{R}\right)^n\right) \qquad \text{(E.1)}$$

a. Show that the mean flow velocity, averaged over the cross-sectional area, is given by:

$$U = U_{max}\left(\frac{n}{n+2}\right) \qquad \text{(E.2)}$$

b. Hence show that the frictional force per unit length of the vessel is given by:

$$f = -2\pi\mu(n+2)U \qquad \text{(E.3)}$$

This result means that we have just three variables remaining. The usual way to complete this set of equations is to consider the relationship between the fluid pressure and the vessel wall area, along the lines that we derived earlier. Re-writing Eq. 8.13 in terms of vessel area gives:

$$p - p_o = G_o\left(\sqrt{\frac{A}{A_o}} - 1\right) \qquad \text{(8.16)}$$

The governing equations can thus be written as:

$$\frac{\partial A}{\partial t} + \frac{\partial Q}{\partial x} = 0 \qquad \text{(8.17)}$$

$$\frac{\partial Q}{\partial t} + \frac{\partial}{\partial x}\left(\frac{Q^2}{A}\right) + \frac{G_o}{2\rho}\sqrt{\frac{A}{A_o}}\frac{\partial A}{\partial x} = -\frac{2\pi\mu(n+2)Q}{\rho A} \qquad \text{(8.18)}$$

where we are writing them in terms of area and flow rate (defined as $Q = UA$), rather than area and velocity.

Now that we have two simultaneous equations, we can examine how they behave. You will notice that they are very similar in form and so it is common to write them in vector/matrix form:

$$\frac{\partial \mathbf{U}}{\partial t} + \mathbf{H}(\mathbf{U})\frac{\partial \mathbf{U}}{\partial x} = \mathbf{B}(\mathbf{U}) \qquad \text{(8.19)}$$

where:

8

$$\mathbf{U} = \begin{bmatrix} A \\ Q \end{bmatrix} \tag{8.20}$$

$$\mathbf{H}(\mathbf{U}) = \begin{bmatrix} 0 & 1 \\ -\frac{Q^2}{A^2} + \frac{G_o}{2\rho}\sqrt{\frac{A}{A_o}} & 2\frac{Q}{A} \end{bmatrix} \tag{8.21}$$

$$\mathbf{B}(\mathbf{U}) = \begin{bmatrix} 0 \\ -\frac{2\pi\mu(n+2)Q}{\rho A} \end{bmatrix} \tag{8.22}$$

These are called the quasi-linear matrix form of the equations and might remind you of the **wave equation**. The eigenvalues of $\mathbf{H}(\mathbf{U})$ can be shown to be the mean flow speed plus or minus the wave speed, and are given by:

$$\lambda = \frac{Q}{A} \pm \sqrt{\frac{G_o}{2\rho}\left(\frac{A}{A_o}\right)^{\frac{1}{4}}} \tag{8.23}$$

The first term is the mean speed of the fluid and the second term is the wave speed. Any disturbance in the flow (such as a pulse of blood) will thus cause a movement both forwards and backwards. This movement will have a speed that is proportional to the square root of the vessel wall stiffness: hence in very stiff vessels, the wave speed is very high. Typical values are that the vessel wall stiffness is of order 10 kPa and the density of blood is about 1000 kg/m^3, so the **pulse wave velocity** is a few metres per second.

Arterial Stiffness

In the limit as the vessel walls become infinitely stiff, the pulse wave speed tends to infinity and we can get shock waves forming. This is one of the reasons why cardiovascular problems develop with age, as vessel walls become stiffer (for example as the result of fatty deposits on the vessel wall). The wave speed will also be higher when the vessel is dilated, i.e. when the pressure is higher, but this is a smaller effect. Measurements of pulse wave velocity, for example using photoplethysmography, can thus be a useful way of measuring the behaviour of the vascular system.

8.4 Blood Pressure

The body's pulse can easily be felt at the wrist by pressing your fingers against the inside of your wrist. This is simply caused by the regular beating of the heart forcing an artery to expand and contract rhythmically as pulses of blood pass through the artery. When we talk about blood pressure, we conventionally mean **Arterial Blood Pressure (ABP)**, which is the pressure found in the arteries and their branches. This pressure varies between a high pressure, caused by ventricular contraction, which is termed **systolic**, as it occurs during systole, and a corresponding low pressure, termed

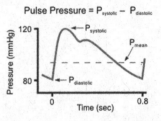

◘ Fig. 8.5 Arterial blood pressure waveform (this figure is taken, without changes, from OpenStax College under license: ▶ http://creativecommons.org/licenses/by/3.0/)

diastolic, as it occurs during diastole. An example waveform is shown in ◘ Fig. 8.5: note that the rise is much sharper than the subsequent fall. Also visible is the 'dicrotic notch', which is the sudden drop and rise in ABP that occurs when the aortic valve closes (see the ▶ Chap. 7 for details). Since there is a peak in blood pressure every time the heart beats, heart rate can be calculated from the blood pressure, in a similar manner to methods used for the ECG.

ABP is nearly always measured in units of millimetres of mercury (mmHg), rather than in Pascals, since the resulting numbers are around 100, which makes them easier to quote (the conversion factor is 1 mmHg = 133 Pa). Normally, only the systolic and diastolic values of blood pressure are measured for each heart beat and this is then recorded in the form systolic over diastolic (for example a systolic blood pressure of 120 mmHg and a diastolic blood pressure of 85 mmHg would be quoted as 120/85).

Mean arterial blood pressure (MAP) is also of interest, since it gives the -average effective pressure that drives blood through the systemic circulation. Although it should strictly be calculated by integrating the blood pressure waveform over time, often only the systolic and diastolic values are known. The most common weighted combination of the two is:

$$MAP = DBP + \frac{SBP - DBP}{3} \tag{8.24}$$

where SBP and DBP represent systolic and diastolic blood pressures respectively. Although it would seem more logical simply to average the SBP and DBP, it turns out that for a typical waveform, this weighting is more representative of the time-averaged value.

There are some other relationships that we use to help to understand the flow of blood through the body. We have already come across Cardiac Output and the concept of resistance to flow, so we introduce the idea of **Total Peripheral Resistance**:

$$TPR = \frac{MAP}{CO} \tag{8.25}$$

which is essentially Eq. 8.3 applied to the whole body. Of course in this equation we are assuming that the blood pressure at return to the heart is zero. Although very simple, Eq. 8.25 does tell us that to change MAP the body must either change CO or TPR. We will look at both of these later.

We also define the **Arterial Pulse Pressure** (APP) to be:

$$APP = SBP - DBP \tag{8.26}$$

i.e. the difference between the maximum and minimum value of arterial pressure during each heartbeat. It turns out that there is an approximate relationship between APP and heart stroke volume:

$$APP = \frac{SV}{C_a} \tag{8.27}$$

where C_a is arterial compliance.

Changes in APP are thus caused either by a change in heart stroke volume or arterial compliance. It can be seen that MAP and APP are in many ways more useful measures of how the cardiovascular system is behaving than SBP and DBP.

Blood flow through the body has one predominant frequency, which is the heart rate (although there are actually many other frequency components to a blood pressure signal, caused by respiration and other processes within the human body). The equivalent circuit shown in ◘ Fig. 8.4 has several sections that can be thought of as filters, whereby frequency components are removed.

A resistor and inductor in series have a time constant:

$$\tau = \frac{I}{\mathcal{R}} = \frac{\rho L}{\pi R^2} \cdot \frac{\pi R^4}{8\mu L} = \frac{\rho R^2}{8\mu} \tag{8.28}$$

If we assume that the dominant frequency is 1 Hz (the heart beat), then for the inductor to be important, we need:

$$\omega > \frac{1}{\tau} \tag{8.29}$$

i.e. the frequency of oscillation must be greater than the cut-off frequency. This will be true when:

$$R > \sqrt{\frac{4\mu}{\pi\rho}} \tag{8.30}$$

Since blood has roughly the same density as water and three times the viscosity, this turns out to be approximately 2 mm. Inductance is thus only important in vessels above this size (which is why we only included it in ◘ Fig. 8.4 for the arterial and venous compartments).

A resistor and capacitor have time constant:

$$\tau = \mathcal{R}C = \frac{12\mu L^2}{ERh} \tag{8.31}$$

This will be important in long vessels with thin walls and low values of Young's modulus. This makes it most important for the venous compartment. In fact arterial compliance also turns out to be very useful in the control of blood flow (as we will see in ▶ Sect. 8.6), so this is normally retained.

> **Exercise H**
> For the circuit shown in ◻ Fig. 8.4, derive a simplified expression for the transfer function of the circuit, neglecting inductance. What is the time constant and what condition must be satisfied if this circuit is to remove oscillations in capillary flow due to pulsatile ABP?

8.4.1 Long-Term Measurement Techniques

Measurements of blood pressure are very important to doctors when assessing the risks of a whole range of vascular diseases, for example strokes, where the risk of a stroke goes up rapidly with increased blood pressure. Elevated blood pressure (**hypertension**) will normally be treated through the use of anti-hypertensive drugs that act to lower MAP.

Blood pressure is normally measured using a **sphygmomanometer**, which consists of an inflatable cuff that is placed around the upper arm over the brachial artery and connected to a pressure gauge. The pressure gauge is usually a column of mercury. The cuff is inflated to a pressure well above the systolic pressure (in the region 175–200 mmHg): since this is higher than the largest arterial pressure, the blood vessels collapse, preventing blood flow to or from the forearm. A stethoscope is then placed over the artery just below the cuff and the cuff pressure allowed to fall gradually: as soon as the cuff pressure falls below the peak arterial pressure, some blood passes through the arteries, but only intermittently. This gives the value of SBP.

Since this flow is turbulent and intermittent, tapping sounds, known as Korotkoff sounds, are produced. As the pressure continues to drop, the sounds increase in volume before decreasing again. The pressure at which the sounds disappear completely is the DBP. Since these sounds are often difficult to hear near the diastolic pressure, accurate determination of the DBP is often a matter of experience. This technique is known as the **auscultatory** technique, since it is based on listening to sounds, and is illustrated in ◻ Fig. 8.6.

It is also possible to use oscillometry to measure blood pressure. This is performed using a cuff with an in-line pressure sensor. Again, the cuff is pressurised above SBP and allowed to deflate: the measured cuff pressure is high-pass-filtered above 1 Hz to observe the pulsatile oscillations as the cuff deflates. The point of maximum oscillation corresponds to a cuff pressure equivalent to MAP. Both SBP and DBP are found where the amplitude of the oscillations is a fixed percentage of the maximum amplitude: 55% for SBP and 85% for DBP.

8.4.2 Short-Term Measurement Techniques

We should note that both of these methods have a measurement time that is much longer than any individual heart beat period: they are thus used for occasional samples rather than as continuous measures. Some attempts have been made to provide a continuous measure of blood pressure. The continuous vascular unloading technique

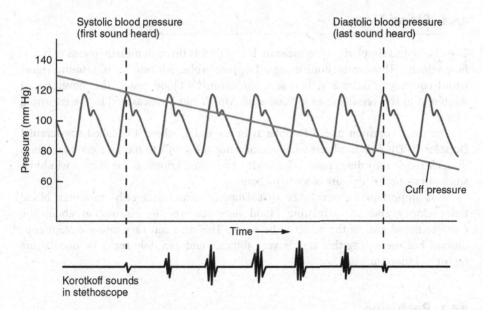

Fig. 8.6 Blood pressure measurement (this figure is taken, without changes, from OpenStax College under license: ▶ http://creativecommons.org/licenses/by/3.0/)

works on the basis of continuously adjusting the cuff pressure to the arterial pressure: this is done using mechanical feedback. The main disadvantage of this technique is that it can only be applied at a peripheral vascular location such as the finger, where it may not be an accurate representation of true arterial blood pressure. There are a variety of other possible techniques, each with advantages and disadvantages, which we will not consider here.

8.5 Measurement of Blood Supply

As well as measuring blood pressure, we are often interested in measuring blood flow to different parts of the body. Blood carries nutrients to body organs and so blood supply must be provided continuously, particularly to organs such as the brain and heart. Cardiac output is adjusted to provide sufficient total blood flow and the resistances of different body organs are also altered to ensure that each organ receives enough flow.

An important function of the circulation is to support metabolism by supplying nutrients and removing waste. Thus we are often interested at the capillary level less in blood flow and more in terms of blood supply or **perfusion**. Perfusion measures the rate of delivery of blood to a volume of tissue, rather than simply the flow rate through the vessels; thus it has units of volume of blood per volume of tissue per unit time, typically ml blood/ml tissue min^{-1}. As we will see in ▶ Chap. 9, the delivery and the removal of gases are essentially perfusion limited processes, because these predominantly happen over the large surface area of the capillary bed, i.e. it is perfusion that determines how much is delivered to and taken away from tissue.

8.5.1 Blood Flow

The easiest and simplest way to measure blood flow is through measurements of blood flow velocity. This can be done using a Doppler probe, whereby an ultrasound signal with frequency of order 2 MHz is sent out towards a blood vessel. The movement of the blood in the vessel causes a phase shift, which can be measured in the returning signal.

The most common use of Doppler is in the brain, where it is called **transcranial Doppler (TCD)**. This has been widely used since the 1980s to monitor cerebral blood flow, although recordings have to be made in what are known as insonation windows, since ultrasound signals are blocked by bone.

It is important to remember that Doppler ultrasound only measures blood flow velocity and so calculating blood flow requires an assumption about the cross-sectional area of the vessel to be made. This area can vary under certain conditions, but measuring this area is very difficult and can only really be done using high-resolution imaging.

8

8.5.2 Perfusion

Perfusion is a key measure of how many organs behave. Hypoperfusion (a drop in perfusion) can have a very detrimental impact on organs such as the heart and brain. In the brain, hypoperfusion can be the result of either a blockage in a supply vessel or a bleed in a major blood vessel in the brain: these cause an **ischaemic** stroke or a **haemorrhagic** stroke respectively. In the heart, hypoperfusion in the blood vessels supplying heart muscle can cause a heart attack, as described in the previous chapter.

Since perfusion is so important to body organs, there has been a lot of interest in measuring perfusion, mainly using imaging techniques to generate perfusion maps. The common factor to all of these is the use of a **contrast agent** that acts as a tracer: the basic principle is that flow is measured by tracking a substance that moves and the concentration of which can be followed. Two very common examples are **Positron Emission Tomography (PET)** and **Magnetic Resonance Imaging (MRI)**.

In PET, a radioactive tracer is injected into the bloodstream: this is most commonly a sugar called F-18 labelled flurodeoxyglucose (FDG). This gradually accumulates in the organ of interest before it disperses away and the radioactivity decays below a detectable level: a time series of the concentration can then be used to estimate perfusion (see ► Chap. 5). In MRI, the most common tracers are based on gadolinium, which is a material whose magnetic properties mean that its presence in tissue alters the MRI image in direct proportion to its concentration.

MRI can also use the water in blood as a naturally occurring, or **endogenous** tracer. This is most often done in the brain, when the magnetism of blood water is inverted in the neck before imaging in the brain. This process, known as Arterial Spin Labelling (ASL) allows water accumulation in tissue to be seen, by which time it has had time to exchange from the blood into the tissue, so that perfusion can be measured. This technique avoids the need for injections, but does then give a poorer signal to noise ratio.

8.6 Control of Blood Flow

The fact that body organs need a continuous supply of blood means that the body has a variety of mechanisms to maintain this blood supply. We looked at the ways in which the central control mechanisms control blood flow to an organ in the previous chapter: we now consider how individual vessels locally control perfusion. This happens primarily within an organ at the arteriolar level (remember that these are the vessels with the most smooth muscle and the greatest contribution to resistance), although there is thought to be some control at the capillary level too.

Individual Vessels

In Exercise I you will show that, for fixed compliance, flow is proportional to vessel radius to the power 6 and that shear stress is proportional to vessel radius to the power 3. These mean that the vessel is extremely sensitive to even quite small changes in vessel radius. In fact, the vessel wall is thought to respond to a large number of stimuli, including shear stress, and these combine to give a signal that acts to control blood flow. Precisely how this works is not well understood, although it is known that nitric oxide and intracellular calcium play important roles in the behaviour.

Exercise I

Consider a vessel of fixed length with driving pressure P at the inlet and zero pressure both at the outlet and outside the vessel.

a. If the vessel compliance is constant, show that flow through the vessel is proportional to vessel radius to the power 6.
b. Show also that shear stress is proportional to vessel radius to the power 3.

In fact, vessel compliance changes as the vessel alters in size. In Exercise J you will show that for an incompressible vessel, the vessel wall gets thinner as the vessel expands. Thus as pressure increases and the vessel wall gets thinner, compliance increases passively very rapidly (the numerator gets larger and the denominator gets smaller).

Exercise J

a. Consider a vessel fixed at both ends, initially with inner radius R_o and wall thickness h_o. Assuming that the volume of the vessel wall does not change when the pressure increases, derive a relationship for the wall thickness as a function of the inner radius and the initial inner radius and wall thickness.
b. Using the expression for resistance (Eq. 8.1), explain how resistance changes as pressure (and hence inner radius) increases.
c. Does the vessel wall become stiffer or more compliant as pressure increases?

However, in reality vessels become less compliant with increased diameter due to the active mechanisms involved in autoregulation of blood flow. The final response of a blood vessel to a decrease in blood pressure is thus made up of two components: the early passive response that leads to a decrease in diameter and a decrease in blood flow; and the subsequent active response that counterbalances this response to maintain blood flow nearly constant.

Individual Body Organs

One of the most tightly regulated body organs is the brain, where perfusion (known here as cerebral blood flow) is kept nearly constant within a range of ABP of approximately 50–150 mmHg. It does this by tightly controlling blood flow through the arterioles, where there are a lot of smooth muscle cells. Precisely how this is done is still not entirely fully understood, but impaired autoregulation in the brain is found in a wide range of brain diseases, such as stroke, dementia and brain injury.

Body organs, of course, do not just regulate their blood supply in response to pressure, but must also match blood (and hence nutrient) supply to metabolic demands. If the cells need greater oxygen, more blood must be supplied. In the brain this will only be a relatively small change, but in some organs, such as skeletal muscle, the increase in blood supply can be very large. The balance between these two can be modelled very simply.

Let's consider a body organ with driving pressure difference P and metabolic rate of oxygen M. The equations governing flow and metabolism can be written as follows:

$$(C_a - C_v)Q = M \tag{8.32}$$

$$P = RQ \tag{8.33}$$

where C_a and C_v are arterial and venous oxygen concentrations. The first equation comes from conservation of mass (remember cardiac output in the previous chapter) and the second one is simply Eq. 8.3.

We will assume to start with that resistance responds to changes in venous oxygen concentration linearly, so that a decrease in venous oxygen concentration results in a decrease in resistance:

$$R = R_o(1 + AC_v) \tag{8.34}$$

where A is a constant (rather like the gain in a feedback circuit). The flow through the organ is then given by:

$$Q = \frac{1}{1 + AC_a}\left(MA + \frac{P}{R_o}\right) \tag{8.35}$$

Obviously if $A = 0$, there is no regulation and flow is linearly proportional to pressure. However, as A increases, there is greater feedback and flow becomes less sensitive to pressure.

If we reference everything back to baseline conditions (denoted by the overbar), then we can re-write the equation for flow as:

$$\frac{\dot{Q}}{\bar{Q}} = \alpha \frac{\dot{M}}{\bar{M}} + (1 - \alpha)\frac{\dot{P}}{\bar{P}} \tag{8.36}$$

where:

$$\alpha = \frac{A\bar{M}R_o}{\bar{P} + A\bar{M}R_o} \tag{8.37}$$

This means that flow responds to both changes in metabolism (with sensitivity α) and to blood pressure (with sensitivity $1 - \alpha$). There is a trade-off between these two, with it being desirable to have higher sensitivity to changes in metabolism than to changes in pressure. This is achieved by having:

$$A \gg \frac{\bar{P}}{\bar{M}R_o} \tag{8.38}$$

i.e. a high level of feedback. However, any system with too much feedback can become unstable, so it is not as simple as aiming for as high a value as possible. This again is a trade-off between competing requirements.

Whole Body

Now that we have looked at the whole cardiovascular system, we will finish by considering the whole body and how the heart plays its part in controlling blood pressure. We divide the body into arterial and venous compartments, each of which has resistance and compliance. Assuming that both compliances are constant, we can state that total blood volume comprises arterial and venous compartments:

$$V_T = P_a C_a + P_v C_v \tag{8.39}$$

In the steady state we can also say that total resistance is related to pressure and flow by:

$$P_a - P_v = QR_T \tag{8.40}$$

This time, in order to maintain arterial blood pressure, we assume that cardiac output, Q, is proportional to heart rate, H, and venous pressure:

$$Q = kHP_v \tag{8.41}$$

This is a simple approximation, based on the ability of the heart to provide greater cardiac output based on heart rate and preload.

From these, we can derive an expression for arterial pressure:

$$P_a = \frac{V_T(1 + kHR_T)}{C_v + C_a(1 + kHR_T)} \tag{8.42}$$

At low values of heart rate, this approximates to:

$$P_a = \frac{V_T}{C_v + C_a} \tag{8.43}$$

but at high values of heart rate, this approximates to:

$$P_a = \frac{V_T}{C_a} \tag{8.44}$$

which is larger. However, this is the maximum arterial blood pressure than can be achieved in this model and so continuing to increase heart rate to raise arterial blood pressure becomes less effective. The model also explains why as blood vessels become stiffer with age (and compliance decreases) baseline arterial blood pressure tends to increase.

We can plot this result, which is most easily done by converting it to non-dimensional form:

$$\frac{P_a}{\overline{P}_a} = \left(1 + \frac{1}{p_r c_r}\right) \frac{(1 + h(p_r - 1))}{\left(1 + h(p_r - 1) + \frac{1}{c_r}\right)} \tag{8.45}$$

where arterial blood pressure, as a fraction of its baseline value, is only a function of heart rate (now as a fraction of its baseline value), h, and the ratio of arterial to venous blood pressure, p_r, and the ratio of arterial to venous compliance, c_r. By writing the result with as few variables as possible in non-dimensional form, we can see that there are only two extra parameters determining the relationship between heart rate and blood pressure, making it much easier to interpret. The resulting relationship between heart rate and blood pressure is then shown in ◘ Fig. 8.7, where we have taken $p_r = 10$ and $c_r = 1/3$ as typical values. The saturation is very clearly seen, where increases in heart rate have progressively less impact on arterial blood pressure.

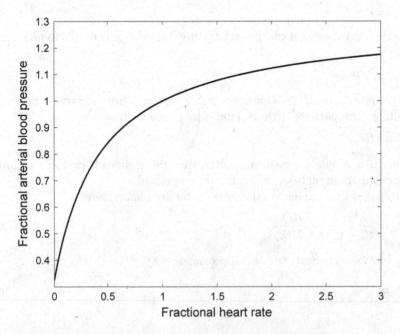

◘ **Fig. 8.7** Relationship between heart rate and blood pressure in simple model

In these two examples, we can see both how whole organ control is achieved through very simple feedback on oxygen concentration and whole body control through blood pressure and heart rate. Obviously the models that we are using are very simplistic, but they do illustrate the kind of behaviour that is seen in the human body and help to provide an insight into the underlying relationships.

8.7 Conclusion

In this chapter, we have examined the second part of the cardiovascular system: all the vessels that make up the vasculature. These can be modelled using very simple techniques and we have developed the equivalent electrical circuit model that is in very widespread use. We have examined how these models can be constructed and developed. We then investigated how blood flow and perfusion can be measured and finished by considering how the body regulates blood flow at a local level in addition to the global control that we looked at in ▶ Chap. 7.

Strokes

We discussed in ▶ Chap. 7 what can happen when there is a drop in blood flow to the heart (i.e. a heart attack). A similar drop in blood flow to the brain due to a blocked blood vessel results in an ischaemic stroke (sometimes called a 'brain attack'). This results in the death of cells in the tissue that was supplied by this particular vessel.

If the clot that causes the blockage is large and affects a large amount of brain tissue, it will need to be removed very quickly (typically within around 4–5 h) otherwise the person having the stroke will be very badly affected. Typical results of a severe stroke can be a loss of speech, a loss of mobility and other disabilities, often requiring long term care.

The Respiratory System

Contents

© Springer Nature Switzerland AG 2020
M. Chappell and S. Payne, *Physiology for Engineers*, Biosystems & Biorobotics 24,
https://doi.org/10.1007/978-3-030-39705-0_9

Having considered in ▶ Chap. 8 the circulatory system, which is responsible for 'bulk' transport around the body, we now move to the respiratory system to understand how gases get in and out of the body. We will also consider more of the details of the gas transport process itself, as this is a critical part of homeostasis.

9.1 Introduction

The respiratory system can roughly be divided into the upper airways in the head and neck, and the lower airways including trachea and all the structures in the lungs. The primary functions of the respiratory system are:

- Gas exchange—most importantly the movement of O_2 into the body and CO_2 out.
- Host defence—it provides one of the barriers between the outside world and the body inside.
- Synthesis and metabolism of various compounds.

9.2 The Lungs and Pulmonary Circulation

The lungs are contained in space with a volume of approximately 4 L but present a surface area of approximately 85 m² for gas exchange. This is achieved through a highly branched structure, ◘ Table 9.1, the trachea branching into the two main stem **bronchi** which themselves divide into further bronchi at progressively small diameters. Subsequently further generations of the branches are called **bronchioles** before finally terminating in the **alveoli**.

Both the smallest bronchioles and alveoli are involved in respiration over a gas-blood barrier that is only 1–2 μm thick. The pulmonary capillary bed is the largest in the body having a surface area of approximately 70–80 m². The capillary volume of the lungs is 70 ml at rest, but this can increase up to 200 ml during exercise. This is achieved through the recruitment of closed vessels or compressed capillary segments from an increase in pulmonary pressure when cardiac output is increased, along with an enlargement of the capillaries with a rise in internal pressure when the lungs fill with blood.

9.2.1 Breathing

◘ Figure 9.1 shows the breathing cycle. This is mainly achieved by the action of the diaphragm, although the external intercostal and scalene muscles in the chest also play a role. Air is drawn into the lungs by an increase in the chest cavity, pushing the abdominal content downwards. Exhalation is passive during normal breathing, but may involve active effort by muscles during exercise and hyperventilation.

The total lung capacity is the total volume of air that can be contained within the lung. Various other volumes and capacities can be defined as shown in ◘ Fig. 9.2,

◻ Table 9.1 Branching structure of the lungs

	Generation		Diameter (cm)	Length (cm)	Number	Total cross-sectional area (cm²)
Conducting zone	Trachea	0	1.80	12.0	1	2.54
		1	1.22	4.8	2	2.33
		2	0.83	1.9	4	2.13
		3	0.56	0.8	8	2.00
	Bronchioles	4	0.45	1.3	16	2.48
		5	0.35	1.07	32	3.11
		...	...	...	...	...
	Terminal bronchioles	16	0.06	0.17	5×10^4	180.0
Transitional and respiratory zones	Respiratory bronchioles	17	...	...	...	...
		18				
		19	0.05	0.10	5×10^5	10^3
	Alveolar ducts	20	...	...	...	...
		21				
		22				
	Alveolar sacs	23	0.04	0.05	8×10^5	10^4

Following Weibel ER. Morphometry of the Human Lung. Springer Verlag and Academic Press, Heidelberg, New York 1963

where capacities are defined by combining two or more volumes. Note that under normal breathing conditions, only a relatively small range of the available lung volume is used.

9.2.2 Respiration Rate

We can quantify breathing in terms of the **respiration rate**: this is the number of breaths in a unit time (normally a minute). The measurement of respiration rate at its simplest, therefore, relies on counting the number of breaths per minute. This would typically require either a sensor at the mouth and/or nose to detect air movement of air in and out of the lungs, or a band around the chest. The latter is potentially able to measure changes in chest volume and thus to infer information about lung volume. Alternatively, it can be possible to extract information about respiration from the ECG, since changes in the volume of the abdominal cavity will alter its conductivity and thus produce small fluctuations in the ECG signal. These changes are typically very subtle though and thus hard to observe in anything other than ideal conditions.

9

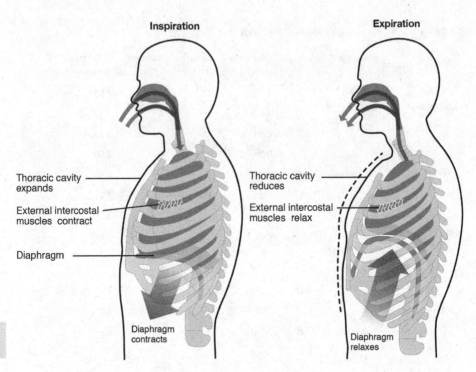

☐ **Fig. 9.1** The breathing cycle (this figure is taken, without changes, from OpenStax College under license: ▶ http://creativecommons.org/licenses/by/3.0/)

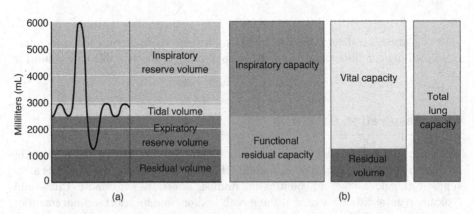

☐ **Fig. 9.2** Definition of the various lung volumes (**a**) and capacities (**b**) (this figure is taken, without changes, from OpenStax College under license: ▶ http://creativecommons.org/licenses/by/3.0/)

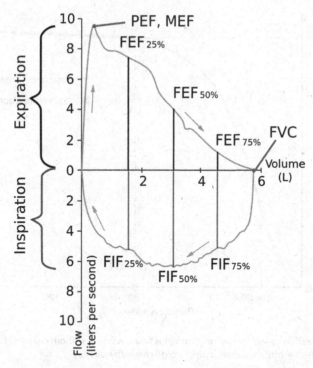

Fig. 9.3 A pneumotachograph, or flow-volume loop, from a forced vital capacity (FVC) test measured using a spirograph. Shown on the loop are the maximal of peak expiratory flow (MEF or PEF), and forced expiratory and inspiratory flow at typical volume fractions (FEF and FIF) (this figure is taken, without changes, from Wikimedia Commons author SPhotographer, Jmarchn, under CC BY-SA 3.0)

9.2.3 Spirometry

Spirometry is a method to measure the volume and flow of air that can be inhaled and exhaled. It is one of a number of pulmonary function tests, and is the most common one used clinically. A spirometer is a device that measures the flow rate of breath passing through it: the patient breathes into a mouth piece, and using that information the volume of air that has passed can then be calculated. Spirometry can generate a pneumotachograph, like the one in ■ Fig. 9.3, which plots the volume and flow of air during a single cycle of inhalation and exhalation.

The most basic spirometric test is one of forced vital capacity (FVC), where a person is asked to take the deepest breath they can and exhale into the sensor as fast as possible, followed by a rapid inhalation. This would give rise to the flow-volume loop in ■ Fig. 9.3. Additionally, the volume-time curve might be produced and examined. From the flow-volume loop it is possible to measure the forced expiratory flow (FEF), which is the flow (in litres per second) defined at discrete fractions of the FVC. Likewise, the forced inspiratory flow can also be measured. A useful simple

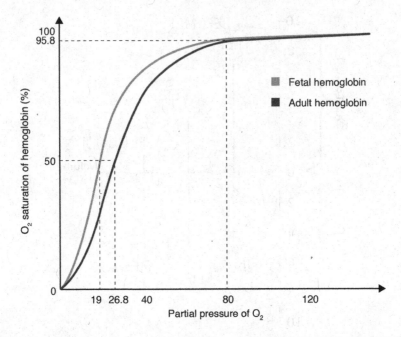

Fig. 9.4 Saturation curves for haemoglobin (this figure is taken, without changes, from David Iberri under license: ▶ http://creativecommons.org/licenses/by-sa/3.0/)

measure is the maximal or peak expiratory flow (MEF or PEF). In principle, MEF and PEF are identical, but it is common to measure PEF using a 'peak flow' meter that is a much simpler device that specifically only measures the peak flow rate. Various other flow, volume and timing parameters can be extracted using spirometry under different breathing conditions, allowing detailed assessment of lung function where required.

Reduced airflow and Chronic Obstructive Pulmonary Disease

A key feature of the lungs, as we have seen in this section, is presenting a large surface area over which gas exchange can occur. This is important so that the body can maintain a sufficiently rapid rate of uptake of oxygen to feed body tissues and the removal of waste carbon dioxide from body tissues. Chronic Obstructive Pulmonary Disease (COPD) is the most common type of obstructive lung diseases. In COPD, airflow in the lungs is limited through the combination of breakdown of the walls of the alveoli, with these spaces being lost to gas exchange, and obstruction of the smaller bronchioles.

Spirometry is often used to diagnose and determine the severity of COPD. Two parameters are measured: the Forced Expiratory Volume in one second (FEV_1), i.e. how much air can be breathed out by the patient in one second; and the Forced Vital Capacity (FVC), i.e. the greatest volume air that can be breathed out in a single breath. Typically, the ratio of these two quantities, FEV_1/FVC would be 75–80%, and a value of less than 70% might indicate that the person has COPD in the presence of other symptoms.

9.3 Gas Transport

Air is a mixture of various gases, predominantly nitrogen (78%) with a large proportion of oxygen (21%) and various other gases (1%) including carbon dioxide. We often refer to the partial pressures of each of the components, where the partial pressure of gas i is defined by:

$$p_i = y_i p, \tag{9.1}$$

where y_i is the mole fraction of the gas and p is the total pressure of the gas mixture. Here the partial pressure is the pressure that this gas would exert if it alone were present. Since y_i represents a fraction of the total that one gas comprises within a mixture, by definition:

$$\sum_i y_i = 1 \tag{9.2}$$

Since in the lungs gases are (indirectly) in contact with a liquid, the blood, we need to be able to relate partial pressure of a gas to its concentration in the blood. We can do this via Henry's law:

$$c_i = \sigma_i p_i, \tag{9.3}$$

where σ is the Ostwald solubility co-efficient. Typical values for respiratory gases in plasma are given in ◘ Table 9.2. All of these solubilities are fairly similar and are in fact too small to carry sufficient quantities of oxygen and carbon dioxide in the blood (hence the need for oxygen to bind to haemoglobin, as we saw in ▶ Chap. 1).

9.3.1 Inert Gases

A simple model for the transfer of gas across the capillary-alveoli interface assumes that the flow is linearly proportional to the difference in partial pressures, thus:

$$q = D_s (c_b - \sigma p_g), \tag{9.4}$$

where q is the net flux per unit area, p_g is the partial pressure of gas in the air space, c_b is the concentration of the gas dissolved in the capillary blood and D_s is the surface diffusion co-efficient. Note that like a lot of the processes we have met in this book, this is just a first order equation with time (since flux is a rate of change of volume with time).

◘ Table 9.2 Solubility of respiratory gases in blood plasma	Substance	σ (mM/mm Hg)
	O_2	1.4×10^{-3}
	CO_2	3.3×10^{-2}
	CO	1.2×10^{-3}
	N_2	7×10^{-4}
	He	4.8×10^{-4}

Exercise A
Starting with the diffusion equation in one dimension, show that the flux of gas per unit area of membrane from the air space into the blood is given by Eq. 9.4, where it has been assumed that the gas has the same solubility in the membrane to that in the blood.

As you will see in Exercise A modelling the flow of gas across the capillary-alveoli interface is very similar to Exercise 4A where we modelled passive transport through a cell membrane. Note that we haven't worried about the relative solubilities of the gas in the membrane compared to the blood here. This is because the membrane that we care about is the cellular layer that separates the gas space and the blood. These cells will have an aqueous interior and thus similar solubility to that of blood. We have ignored the diffusion through the cell membranes (and thus relative solubility of blood in the cell walls) because these are much thinner than the cells themselves.

A complete pulmonary capillary can be modelled as a cylinder with gas transport occurring over the full surface. In the limit the expression for total gas flux across the surface, which is the flow Q, reduces to:

$$Q = F\sigma(p_0 - p_{in}), \tag{9.5}$$

where p_{in} is the partial pressure of the gas in the blood at the inlet to the capillary. Thus the gas flux depends only upon the pressure difference and the blood flow rate, F. This only holds as long as the permeability of the membrane to the gas (i.e. the D_s value) is sufficiently high, which is primarily achieved in the lungs through having a thin membrane. Since the gas flux depends on blood flow and not the rate of diffusion it is called **perfusion limited**.

Exercise B
Here we will derive the result in Eq. 9.5 based on a cylindrical model of a pulmonary capillary.

(a) Using continuity derive the relationship between the change in concentration of gas within the blood along the capillary and the flux across the surface for a vessel of radius r and with a blood flow rate F.

(b) Find the solution to this equation for the concentration of gas within the blood with distance along the capillary, x. Assuming that the concentration at the inlet is c_0 and that the concentration in the air space is constant along the length.

(c) The total gas flux across the capillary surface can be calculated for a capillary of length L by integration. Using the solution from part (b) derive an expression for the total gas flux

(d) Calculate the flux for a single capillary, assuming parameter values of:
$C_{in} - C_o = 1$ mM, $F = 1 \times 10^{-13}$ m³/s and $2\pi r D_s L = 0.51 \times 10^{-13}$ m³/s. Plot the variation of flux with blood flow rate, assuming that all other parameters remain constant.

(e) Using the result in part (d), explain why increasing the flow rate has only a small effect on the total flux across the capillary wall.

(f) What happens in the limit as $L \to \infty$? What properties of the capillary bed structure would ensure that this would be true?

9.3.2 Carbon Dioxide

Whilst the analysis above holds for inert gases, where Henry's law holds, the metabolic gases are more complicated. CO_2 is mainly transported within the red blood cells as HCO_3^-, the reaction being catalysed by the enzyme carbonic anhydrase:

$$CO_2 + H_2O \rightleftarrows H_2CO_3 \rightleftarrows HCO_3^- + H^+, \tag{9.6}$$

where CO_2 combines with water to form carbonic acid, which then produces bicarbonate (HCO_3^-) and H^+.

Exercise C

(a) Determine the rate constant for the second part of the reaction scheme in Eq. 9.6 for CO_2 conversion to carbonic acid.

(b) Convert this to the 'corrected' rate constant assuming that almost all the available CO_2 converts to carbonic acid.

In Exercise C you have derived an expression for the corrected rate constant, K_A, that relates CO_2 concentration to the concentrations of hydrocarbonate ions and H^+. It is possible to evaluate what difference this makes to the removal of CO_2 from the capillary blood in the lungs and arrive at the result:

$$Q = f(1 + K_A)\sigma_{CO_2}(p_0 - p_{CO_2}), \tag{9.7}$$

which is $(1 + K_A)$ larger than our result for inert gases in Eq. 9.6, with $K_A = 20$ at a normal pH of 7.4. The conversion of CO_2 to bicarbonate thus substantially increases the gas flow rate, because as CO_2 leaves the capillary by diffusion from the plasma it is rapidly replenished by more from the reversal of the bicarbonate reaction. Thus this is an example of **facilitated diffusion** that we met in ▶ Chap. 4.

Exercise D

(a) Starting with the model in Exercise B write down separate equations for the concentration of CO_2 and HCO_3^- in the capillary blood. Assume that you can ignore H_2CO_3 from your reaction scheme and thus have a single forward and backward reaction rate.

(b) Combine your equations from part (a) to give a single expression for the total concentration of CO_2 and HCO_3^-, i.e. CO_2 in all its 'forms' for the purposes of transport.

(c) Using a quasi-steady state approximation write HCO_3^- in terms of CO_2 and thus derive the result in Eq. 9.7 using a similar analysis to Exercise B and considering the limit as $L \to \infty$.

(d) If $\log_{10} K_a = -6.1$ calculate K_c at normal pH and thus comment on the effect of blood chemistry on CO_2 removal from the body.

9.3.3 Oxygen

We have already seen in ▶ Chap. 1 that oxygen binds to haemoglobin the blood:

$$Hb + 4O_2 \rightleftarrows Hb(O_2)_4, \tag{9.8}$$

This is the primary means by which it is transported; only a negligible fraction (3%) is carried in solution in the plasma. ◘ Figure 9.4 shows the relationship between the partial pressure of O_2 and haemoglobin saturation, which is highly non-linear. In theory the enhancement provided by haemoglobin on oxygen transport could be as much as 200, but in practice an enhancement of around 32 is achieved. Interestingly, as shown in ◘ Fig. 9.4, the haemoglobin saturation curve is different for fetal blood.

9.3.4 Tissue Gas Delivery

Delivery of gases to the tissues is essentially the reverse of the process in the lungs, and again the process is typically assumed to be perfusion limited. We can thus equate the amount of gas delivered to the difference in partial pressure between the arterial and venous ends of the capillary bed. This process can be modelled by a linear first order differential equation, just as in Eq. 9.5:

$$\frac{dp_t}{dt} = F \frac{\sigma_b}{\sigma_t} (p_a - p_v), \tag{9.9}$$

where p_a, p_a, and p_t are the partial pressure in arterial blood, venous blood and tissue respectively, and we have to consider the solubility of the gas in both the blood and tissue.

This means that gas delivery can be considered to be like the one-compartmental models that we considered when we examined pharmacokinetics in ▶ Chap. 4. In this case the compartment is the tissue into which gas is being delivered and we assume that there is no spatial variation in the partial pressure in the tissue, thus the compartment is referred to as 'well stirred' or 'well mixed'. This can be represented graphically as in ◘ Fig. 9.5, where a number of other simple models of tissue-capillary gas exchange are shown that extend the model by accounting for some degree of **diffusion limitation**.

The one-compartment model means that the tissue concentration will respond exponentially to a change in gas concentration and a time constant can be deter-

Gas exchange model	Graphical representation
a) Perfusion-limited	
b) Perfusion-limited: two parallel compartments	
c) Perfusion-limited counter-current diffusion	
d) Perfusion-diffusion	
e) Perfusion-diffusion counter-current diffusion	

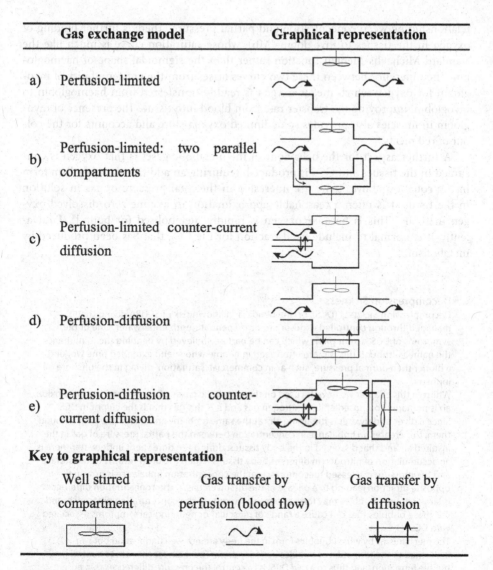

Key to graphical representation

Well stirred compartment	Gas transfer by perfusion (blood flow)	Gas transfer by diffusion

◻ **Fig. 9.5** Compartmental models of tissue gas exchange (reproduced with permission, M.A. Chappell, DPhil thesis, University of Oxford, 2006)

mined for that process; often the half-life is quoted. This will vary from tissue to tissue depending upon the solubility, but more significantly upon the perfusion. Given the range of tissues and accompanying perfusion values, simple models often divide the body into a collection of compartments with a range of representative time constants.

As before, this description is fine for the inert gases, but is insufficient for the metabolic gases. As we have already seen, binding in the blood gives a non-linear relationship between concentration and partial pressure. There is further binding of oxygen in the tissues to myoglobin (Mb), whose saturation curve is much like the standard Michaelis-Menten function rather than the sigmoidal shape of haemoglobin. The difference between these two curves arises from the greater affinity of myoglobin for oxygen which means oxygen is readily transferred from haemoglobin to myoglobin, improving the transfer rate from blood into tissue. The presence of myoglobin in muscles also provides some limited oxygen store and accounts for the colour of red meats.

A further aspect for the behaviour of the metabolic gases is that oxygen is consumed in the tissues and CO_2 is produced, requiring an additional metabolism term in the equation above. If we are interested in the total pressure of gas in solution in the tissue it is often a reasonable approximation to assume zero dissolved oxygen in tissue. This is because oxygen is rapidly metabolised (or bound). Equivalently, it is normal to include fixed fraction for the CO_2 that has been produced by metabolism.

Decompression Sickness

Decompression Sickness (DCS), or 'the bends', is a known risk of SCUBA diving that is managed through controlled exposure to time spent at depth underwater. One of the symptoms of DCS is joint pains which can be partially relieved by bending the joint (hence the name), although this is more often seen in people who spend extended time working a higher than normal pressure, such as in commercial 'saturation' diving in the offshore oil industry.

Whilst in this chapter we have focused on the metabolic gases of oxygen and carbon dioxide, air is primarily composed of inert nitrogen gas and it is this gas that is the culprit in DCS. Since a diver underwater is breathing air* at the same ambient pressure as the water around them, this creates an imbalance in concentration between the partial pressure of gas in the inspired air and that dissolved in the body tissues: during the time spent underwater there is an accumulation of nitrogen in different body tissues. When the diver returns to the surface and the pressure is released (decompression) the concentration imbalance is now in the opposite direction leading to a release of the dissolved gas. If the concentration difference is large enough, bubbles may form as a means to reduce the concentration difference, and it is these bubbles that can cause a range of muscular or neurological symptoms associated with DCS.

The mechanisms by which bubbles form in the body are not well understood but can be likened to the formation of bubbles in a fizzy drink. The main mechanism to control bubble formation, and thus to avoid DCS, is to control the pressure differences that arise during decompression, normally by controlling the ascent rate from depth and inserting 'stops' at shallower depths. The way that the rate and 'stops' are calculated typically uses compartmental models like those we have met in this chapter. These allow the effects of accumulation of nitrogen in different groups of tissues to be modelled based on the time spent at depth during the dive and thus the relevant concentration difference that will occur on decompression predicted.

*It is worth noting that divers carry normally carry air in the SCUBA tank and not as is commonly assumed pure oxygen, since oxygen is toxic at partial pressures much above 1 atmosphere. More advanced diving can use other gas mixtures, partly with the aim of reducing DCS.

9.3.5 **Blood Oxygenation**

The standard measurement of blood oxygenation is the fraction of the carrying capacity of the blood that is being used: this is known as **oxygen saturation**. Pulse-oximetry is a widely used method for measuring the oxygen saturation of the blood. The sensor is placed in a thin part of the body, for example a fingertip or earlobe. It works by passing two distinct wavelengths of light through the body to a photodetector; the relative absorbance at the two wavelengths allows the oxygen saturation of arterial blood to be determined.

The measurement relies upon the changes in colour of haemoglobin with oxygen saturation, thus pulse-oximetry strictly measures the percentage of haemoglobin that is loaded with oxygen. Since, as we have seen, haemoglobin is the main carrier of oxygen in the blood, this is a good measure of oxygen saturation of the blood. ◘ Figure 9.6 shows the absorption spectra for both oxy- and deoxy-haemoglobin.

Note that pulse oximetry works in the Near Infrared Region (NIR) of the light spectrum and exploits the distinctive differences in absorption seen in this region. Typically two wavelengths, 660 and 940 nm, are used and are sampled up to 30 times per second allowing changes in ambient light and also the amount of arterial blood present to be corrected for. Once the ratio of light transmission at the two wavelengths

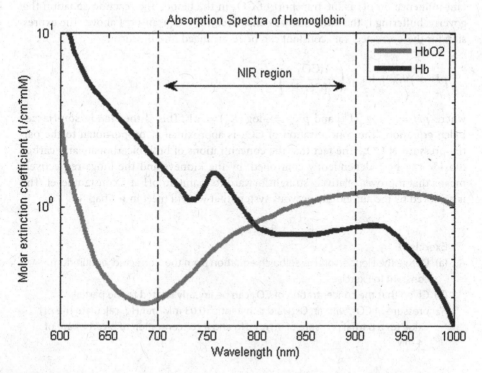

◘ **Fig. 9.6** Absorption spectra for oxy- and deoxy-haemoglobin, showing the Near Infrared Region (this figure is taken, without changes, from Adrian Curtin under license: ▶ http://creativecommons. org/licenses/by-sa/3.0/)

has been calculated this can be converted to oxygen saturation, which in normal healthy individuals will be above 95%.

Whilst pulse oximetry is widely used for the measurement of oxygen saturation, the time course associated with the measurement of optical transmission, the **photoplethysmogram** (PPG), is less widely used. This signal captures the time varying blood volume in the skin associated with the differences between systolic and diastolic pressure in the arteries. Thus, it is possible to monitor heart rate from the PPG. It is also, theoretically, possible to monitor respiration, since changes in lung volume and that of the thoracic cavity affect the heart. This leads to small changes in systolic and diastolic pressures during the breathing cycle that might be detected from the PPG. Measuring such small changes is challenging, but this has not prevented attempts to measure vital signs using the light reflected from exposed areas of skin acquired using conventional digital video cameras.

9.3.6 Control of Acid-Base Balance

One final aspect of the body's behaviour that we will consider here is the regulation of the acid-base status. The maintenance of the pH of arterial blood between 7.35 and 7.45 is absolutely vital for the correct functioning of the human body. The most important influence on pH is the transport of CO_2 in the blood. The reaction equation that governs buffering is the same one we considered for CO_2 transport above. The expression for the 'corrected' rate constant can be re-arranged into the form:

$$pH = pK_A + \log_{10}\left(\frac{[HCO_3^-]}{[CO_2]}\right),$$ (9.10)

where $pH = -\log[H^+]$ and $pK_A = -\log(K_A) = 6.1$. This is the **Henderson-Hasselbalch equation**. The concentration of CO_2 is approximately proportional to the partial pressure of CO_2. The fact that the concentrations of both bicarbonate and carbon dioxide can be independently controlled, by the kidneys and the lungs respectively, means that it proves relatively straightforward to maintain pH at a constant level. This is achieved by the autonomic nervous system that we will meet in ▶ Chap. 10.

Exercise E

(a) Derive the Henderson-Hasselbalch equation from the 'corrected' equilibrium constant in Exercise C.

(b) Given that the concentration of CO_2 can be linearly related to the partial pressure of CO_2, with an Ostwald constant of 0.03 mM/mmHg, calculate the pH value for a partial pressure of 40 mmHg and a concentration of bicarbonate of 24 mM.

(c) Draw lines of constant pCO_2 (20, 40 and 60 mmHg) on a plot of bicarbonate concentration versus pH. Use a range of pH of 7.1–7.7 and bicarbonate concentration of 10–40 mM and illustrate the point calculated in part (b).

(d) If the pCO_2 rises from 40 to 60 mmHg (why might this happen?), how can the body maintain a constant pH?

9.4 Conclusions

In this chapter we have looked at the structure of the respiratory system and especially the respiratory circulation. We have used models of transport from earlier chapters to explain the unique structure of the lungs that enables effective gas exchange. You should now be able to set up compartmental models for gas transport and to analyse these for inert and metabolic gases using suitable approximations.

The Central Nervous System

Contents

© Springer Nature Switzerland AG 2020
M. Chappell and S. Payne, *Physiology for Engineers*, Biosystems & Biorobotics 24,
https://doi.org/10.1007/978-3-030-39705-0_10

We have so far looked at a number of the most vital systems in the body. However, their actions and interactions all need to be coordinated and this role falls to the nervous system, which we will briefly look at in this final chapter.

10.1 Introduction

The nervous system can be divided into two parts:
- The central nervous system (CNS): the brain and spinal cord.
- The peripheral nervous system (PNS): nerves (bundles of nerve cells) connecting the CNS to other parts of the body. The PNS also encompasses the enteric nervous system, a semi-independent part of the nervous system responsible for the gastrointestinal system.

■ Figure 10.1 shows a conceptual model of how this all fits together. Notice that we have both **afferents** (inputs) and **efferents** (outputs), but also a division into parts of which we are consciously aware and have direct control: the **somatic nervous system**; and the part that we are largely unaware of and happens automatically—the **autonomic nervous system**.

10

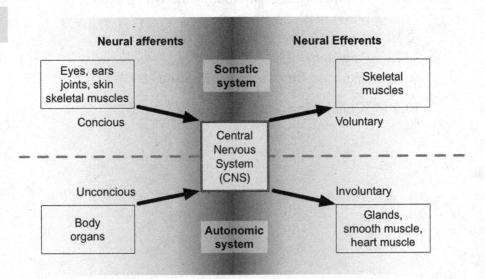

■ **Fig. 10.1**   Conceptual diagram of the nervous system

10.2 Neurons

We have already met the idea of a neuron or nerve cell in ▶ Chaps. 3 and 4 where we considered the action potential and chemical synapses. Neurons are the main actors in the CNS, providing a way to transmit both efferent and afferent signals to and from the CNS. A number of specialized neurons exist:

- Sensory neurons: These respond to stimuli such as light and sound in the sensory organs and send this information back to the CNS.
- Motor neurons: These receive signals from the CNS and cause muscle contraction or affect glands (for the release of hormones).
- Interneurons: These connect neurons to other neurons within the same region of the CNS.

Typically a neuron will have a cell body, dendrites and an axon, as shown in ◘ Fig. 10.2. The dendrites are thin structures arising from the cell body that typically branch multiple times, forming a complex 'dendritic tree'. In contrast the axon (and there is only ever a single axon) is a special extension of the cell body that may extend as far as 1 m. The axon itself may branch hundreds of times before it terminates at the dendrite of another neuron. At the majority of synapses signals are sent from the axon of one neuron to the dendrite of another, although there are exceptions to this rule. It is the axon that carries signals over long distances and is thus insulated with a myelin

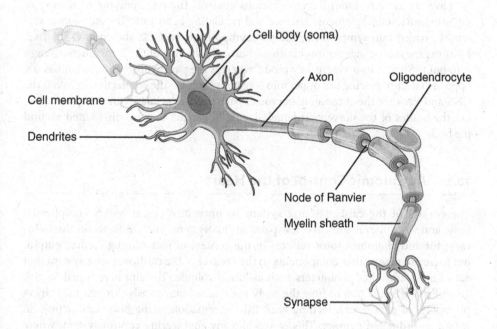

◘ **Fig. 10.2** Schematic of a typical CNS neuron (this figure is taken, without changes, from OpenStax College under license: ▶ http://creativecommons.org/licenses/by/3.0/)

sheath that in turn is interrupted at various points by the nodes of Ranvier to aid conduction of the action potential (AP) as we saw in ▶ Chap. 3.

Neuron Loss and Dementia

Dementia is a broad category of diseases in which sufferers exhibit neurological symptoms associated with loss of neurons in the brain. The most common and well known example is Alzheimer's Disease (AD). A characteristic of AD is the loss of neurons and synapses in the cerebral cortex and certain subcortical regions of the brain. This can be seen in Magnetic Resonance or Positron Emission Tomography imaging as a progressive loss of brain tissue, normally called atrophy, which leads to a shrinkage of the brain. However, seeing atrophy alone is not sufficient to confirm AD or dementia more generally, and some degree of atrophy can be seen in normal elderly brains.

Microscopy studies of brain tissue from people with AD has shown the presence of amyloid plaques, insoluble deposits of beta-amyloid peptides and other cellular material, outside and around neurons. Additionally, neurofibrillary tangles, aggregates of the microtubule-associated protein tau, have been found to have accumulated within the cells. Again, these observations are also made in tissue in healthy individuals as they age, but are observed in higher levels, especially in certain brain regions, in AD. Whether any, or all, of these effects are a result or a cause of AD remains to be established.

10.3 Autonomic Nervous System

10

There are a great many systems in the body that the CNS needs to regulate, over which we have no (or very little) direct conscious control. These include the regulation of digestion, maintaining glucose balance and regulating heart rate. The autonomic system is divided into **sympathetic** and **parasympathetic** parts, as shown in ◙ Fig. 10.3. Most organs receive signals from both and broadly the role of the sympathetic system is to put the organ into 'emergency mode', whereas the parasympathetic system has the opposite effect of placing the organ into 'vegetative mode'. The connection between the CNS and PNS for the autonomic nervous system occurs primarily in the spinal cord, but the bodies of the nerve cells largely reside in ganglia that are distributed around the body.

10.3.1 Autonomic Control of the Heart

The control of the cardiovascular system is, unsurprisingly, a highly complicated topic and we will only look at a few parts of this system. We will focus on the **baroreceptor** and **chemoreceptor** reflexes in the context of maintaining cardiac output, but there are many other components to the control of the cardiovascular system that act to maintain other parameters such as blood volume. The aim here is just to give you a brief introduction to how the body maintains homeostasis through measuring physiological parameters, feeding back this information to the brain and acting on it in a co-ordinated manner. This is just like any engineering control system, where feedback based on measurements is used to control a system and hence to maintain equilibrium.

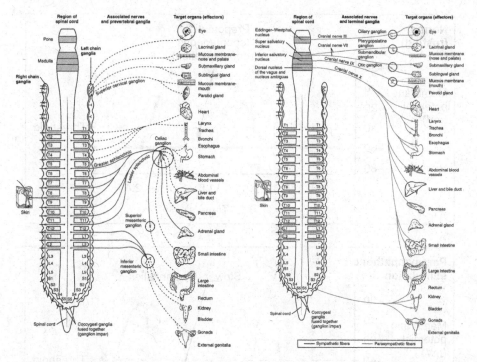

Fig. 10.3 The autonomic system showing sympathetic and parasympathetic parts separately (this figure is taken, without changes, from OpenStax College under license: ▶ http://creativecommons. org/licenses/by/3.0/)

There are three cardiovascular centres in the medulla oblongata in the brain. These neurons respond to changes in blood pressure, blood gas concentrations and pH. The **cardioaccelerator centre** acts to increase cardiac function through sympathetic activation and the **cardioinhibitor centre** acts to reduce cardiac function through parasympathetic activation, whilst the **vasomotor centre** controls vessel wall stiffness via smooth muscle cells.

10.3.2 Cardiac Efferents

The cardiac muscle cells are the main efferents in the heart. As we have seen, their regular contraction is self-sustaining, coordinated by the SAN and requires no external instruction to maintain a resting heart rate. The heart is thus a good example of the competing role of sympathetic and parasympathetic nerves and how these are used to tune the heart rate, as illustrated in ▣ Fig. 10.4:

- Parasympathetic stimulation: Synaptic terminals release the neurotransmitter Acetylcholine (ACh); this slows the rate of depolarization during the pacemaker potential of the SA node, thus increasing the interval between successive action potentials, and slowing the heart rate. ACh acts by increasing potassium

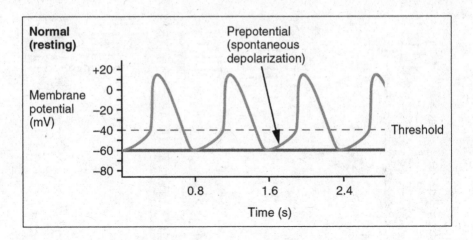

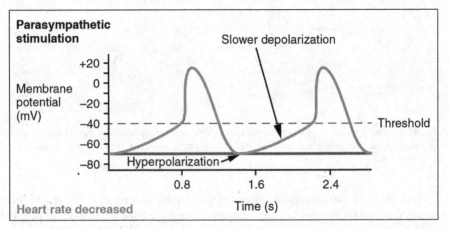

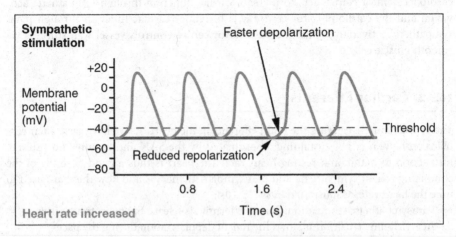

☐ Fig. 10.4 Effects of parasympathetic and sympathetic activity on heart rate (this figure is taken, without changes, from OpenStax College under license: ▶ http://creativecommons.org/licenses/by/3.0/)

10

permeability, keeping the membrane potential nearer to that of potassium, retarding the growth of the pacemaker potential toward the threshold for triggering action potentials.

– Sympathetic stimulation: Synaptic terminals release norepinephrine. This speeds the heart rate and increases the strength of contraction, an effect which is mediated by an increase in calcium permeability. The greater the number of calcium channels that are open, the lower the threshold is for triggering an AP in the SAN, thus increasing the heart rate. In the cardiac muscle cells the increase in calcium permeability increases the calcium influx during the plateau, increasing the strength of contraction.

In both cases the effects of the neurotransmitter are indirect, unlike those we saw when we looked at chemical synapses, where the neurotransmitter acted directly on the ion channels. Note that this means that the body can, via an indirect mechanism, effect longer term changes without having to provide a continuous neural signal.

10.3.3 Cardiac Afferents

The **baroreceptor reflexes** are rapid mechanisms that attempt to minimise short-term fluctuations in blood pressure. Baroreceptors are nerve endings in the walls of the carotid sinus, where the external and internal carotid arteries split, and in the aortic arch. They are termed mechanoreceptors as they are sensitive to stretching of the arterial wall, which is related to arterial blood pressure. When there is a drop in blood pressure, resulting in a reduction in wall stretch, the firing rate of the receptors is reduced.

This feeds back to the brain and results in an increase in heart rate and stroke volume: as a result, there is an increase in blood pressure. The increase in heart rate is known as **tachycardia** (the opposite, a reduction in heart rate, is known as **bradycardia**). How this works is shown in terms of a control loop (with the large arrows used to describe complex relationships) in ◘ Fig. 10.5: a decrease in ABP results in a decrease in baroreceptor firing rate, an increase in cardiac output, which, coupled with **vasoconstriction**, results in an increase of ABP and a return to baseline conditions. There are also many other processes that act to maintain homeostasis in the longer term.

The carotid sinus receptors respond to pressures of approximately 60–180 mmHg, whilst the aortic arch receptors are less sensitive as they have a higher threshold pres-

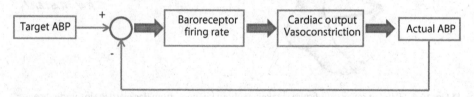

◘ Fig. 10.5 Baroreceptor control of blood pressure

sure. The maximal sensitivity is around normal ABP. The receptors are also sensitive to the rate of change of ABP, and increases in the pulse pressure make the baroreceptors more sensitive to changes in ABP. The baroreceptors quickly adapt to changes in the mean level of ABP: if the level drops for a sustained period, unaltered by the action of the baroreceptors, the reflex will gradually reset itself to this new operating point over a period of several hours. They are thus only short-term regulators of ABP.

The **chemoreceptor reflexes** originate in the aortic arch and carotid bodies and respond to changes in arterial oxygen and carbon dioxide levels. If blood pressure drops and oxygen levels in the blood decrease, the chemoreceptor reflexes will respond by increasing sympathetic activity. This increase acts in the same way as a decrease in parasympathetic activity and so heart rate increases.

10.4 Somatic Nervous System

We will now focus on the somatic nervous system. Remember that this is the part of the CNS whose inputs, neuronal afferents, we are largely conscious of and whose outputs or actions, neuronal efferent, we generally have voluntary control of.

The familiar patellar (knee-cap) reflex provides a good example of the somatic nervous system in action. A simplified diagram of the circuitry involved in this reflex is shown in ◘ Fig. 10.6. Concentrating only on the quadriceps muscle to start with:

10

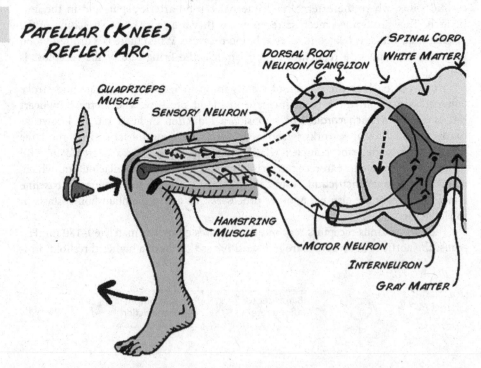

◘ **Fig. 10.6**   Patellar reflex (this figure is taken, without changes, from Backyard Brains under license: ▶ http://creativecommons.org/licenses/by/3.0/)

tapping the knee causes a stretch in the muscle fibres that stimulates an AP in stretch-sensitive sensory neurons. This signal is passed to the spinal cord arriving at a synapse with the motor neuron. This then elicits an AP in the motor neuron, which is carried back to the muscle arriving at a neuromuscular synapse as described in ▶ Chap. 4. The resulting AP causes contraction of the muscle fibre, which when repeated across the whole muscle causes contraction of the quadriceps muscle and the resulting leg extension.

10.4.1 Temporal and Spatial Summation of Synaptic Potentials

The neuromuscular junction is unusual in one respect, in that a single AP that arrives at the presynaptic side leads to a sufficiently large depolarization on the post synaptic side to trigger an AP there. Thus such a synapse is called a 'one-for-one' synapse. However, most synapses are not that strong: a single presynaptic AP will cause only a small depolarization of the postsynaptic cell, called an **excitatory postsynaptic potential** (epsp). The synapse between a single stretch receptor and the quadriceps motor neuron in the patellar reflex is a typical example of this. Each epsp is of the order of 1 mV, far smaller than the 10–20 mV threshold required. However, if several subsequent epsp arrive before the effects of the previous ones have died away a sufficient depolarization of the postsynaptic cell can be achieved. This is called **temporal summation**. An alternative mechanism by which epsp can sum to reach a threshold is via the postsynaptic cell receiving multiple presynaptic inputs, this is called **spatial summation**. This relies on the fact that most neurons are connected to many others via their dendrites. Both temporal and spatial summations are involved in the patellar reflex.

10.4.2 Excitatory and Inhibitory Synapses

So far we have only met excitatory synapses where the AP in the presynaptic cell causes a depolarization in the postsynaptic cell. However, it is also common to find inhibitory synapses, where the release of neurotransmitter due to the presynaptic AP tends to prevent the firing of the postsynaptic AP. In this case the neurotransmitter causes a **hyperpolarization** of the post synaptic cell, making the membrane potential more negative and moving it further way from the threshold required to elicit an AP. Thus we now also have a class of **inhibitory postsynaptic potential** (ipsp). Like the epsp we already met, the ipsp is achieved through changes in the ionic permeability of the postsynaptic cell membrane.

One possible mechanism for an ipsp would be an increase in the permeability to K^+ similar to the undershoot we met in ▶ Chap. 3. Many inhibitory synapses, however, rely on changes in Cl^- permeability. In many neurons chloride pumps maintain the chloride equilibrium potential more negative than the membrane potential, thus an increase in Cl^- permeability leads to a hyperpolarization of the neuron. Even when the Cl^- equilibrium potential is near to that of the membrane potential, inhibition can occur, as, although there will be no appreciable hyperpolarization, the increase in Cl^- permeability will resist any increases in the membrane potential brought about by an excitatory input.

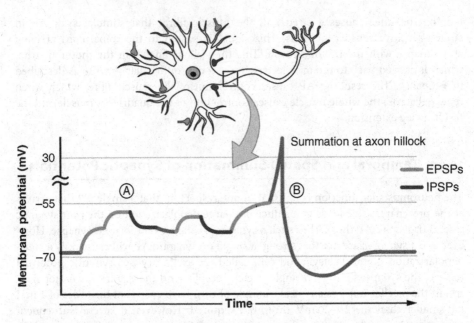

◘ Fig. 10.7   Summation of postsynaptic potentials (this figure is taken, without changes, from Open-Stax College under license: ► http://creativecommons.org/licenses/by/3.0/)

◘ Figure 10.7 shows an example of summation of epsps and ipsps that have been received by a neuron. At time A the sum effect of a series of epsps has been insufficient to initiate an action potential in the axon, despite further ipsps. By time B sufficient epsps have been received within a short enough time to reach the threshold and an action potential will be generated and will propagate along the axon.

The combination of inhibitory and exhibitory synapses plays an important role in the patellar reflex. In ◘ Fig. 10.6 we also have to consider the flexor muscles at the back of the thigh as well as the quadriceps muscle. As the leg extends, this will cause a stretch in these muscles that in turn will, via the excitatory pathway through the spinal cord, cause contraction of the flexor muscle, jerking the leg back again. In turn this would elicit a stretch and contract reaction from the quadriceps muscle and so on. However, the extra inhibitory link between quadriceps sensory neuron and the flexor's motor neuron prevents this occurring and only the first extension is seen.

10.4.3 **The Brain**

The 'pinnacle' of the nervous system is the brain, where many millions of individual neurons interact via a complex network of dendritic connections and synapses. This relies upon the many combinations that are possible when we bring many epsp and ipsp together with temporal and spatial summation over a densely connected neural network.

The brain can broadly be considered to contain both 'grey matter' and 'white matter' reflecting different visual qualities of the tissue, ◘ Fig. 10.8. These actually

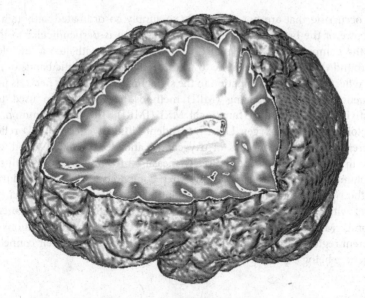

◘ Fig. 10.8 Image of the brain collected using MRI showing folded structure and division into grey and white matters (reproduced with permission, M. A. Chappell)

primarily represent a division of the brain into neurons in the grey matter that forms a thin layer with a large surface area that takes the form of a highly folded sheet within the head, and in the white matter bundles of axons from neurons carrying APs to different regions of the brain and out of the brain. Within the brain there is also another sort of cell called glial cells of which there are a number of different types, all of which provide supporting roles either structurally or metabolically to the neurons.

10.4.4 Function: Introduction to EEG and FMRI

The complexity of the CNS and the brain in particular makes it both very difficult to study and one of the most studied parts of the human body. It is possible to make measurements of neural activity; however, not with a resolution that allows individual cells to be investigated. Like the ECG of the heart that we met in ▶ Chap. 7, **electroencephalography** (EEG) measures electrical signals in the brain using an array of electrodes placed on the head. Alternatively, **magnetoencephalography** (MEG) can be used to detect magnetic fields arising from electrical activity of the brain with highly sensitive magnetic detectors. Both methods provide very highly sampled temporal information about signals in the brain, but provide low spatial resolution, since it is difficult to reconstruct spatial locations from an array of measurements made outside of the head.

Very many EEG electrodes or MEG detectors can be placed across the head to try to localise electrical activity to different areas of the brain. In the heart, the main electrical activity arises from the propagation of the AP across the heart wall, which gives rise to a relatively strong and highly directional vector potential. In the brain many

APs are occurring that are not necessarily so simply co-ordinated, although, due to the structure of the brain, the main direction of travel is perpendicular to the skull. Despite the complexity of the signals in the brain, even with just a few electrodes placed around the head, it is possible to measure signals in specific bands of frequencies that reflect levels of co-ordination in the signalling in the brain's nervous network.

Magnetic Resonance Imaging (MRI) methods have also been used to study brain activity, typically called functional MRI (fMRI). The most common method is to exploit the Blood Oxygen Level Dependent (BOLD) effect. BOLD relies upon the different magnetic properties of oxygenated and deoxygenated blood, allowing increases in oxygen usage to be indirectly measured from changes in perfusion and blood volume in the brain during a task when compared to their values during rest. This method provides far better spatial resolution, of the order of mm, than EEG or MEG, but with much poorer temporal resolution, typically one measurement every few seconds. Much research using these methods has allowed the functions of various different regions of the brain to be identified and increasingly their connections to each other are being explored.

10.5 Conclusions

In this chapter we have examined the nervous system and seen how the concepts of action potentials and their propagation come together to regulate the body as part of the autonomic system and provide for conscious and voluntary action through the somatic system. We have seen that, through the summation of many action potentials, the brain is a very complex organ, requiring in turn a very specific structural architecture and metabolic support system. Finally, despite the complexity of the brain we have seen that we are still able to measure at least some of its functionality by examining electrical and metabolic changes.

At the end of this chapter we have completed our introductory survey of physiology from an engineering perspective. We started with the fundamental unit of living matter, the cell, and examined behaviour associated with it. We then started to examine larger systems where we might bring cells together into the form of a tissue, and then how groups of tissues form systems. Finally, we examined some of the most important systems in the body. You will hopefully have learned a lot about how the human body works, although we have only scratched the surface. More importantly you will have seen that a very wide range of engineering methods can be used to understand and study physiology, bringing interest to engineering and benefit to medicine.

Supplementary Information

© Springer Nature Switzerland AG 2020
M. Chappell and S. Payne, *Physiology for Engineers*, Biosystems & Biorobotics 24,
https://doi.org/10.1007/978-3-030-39705-0

Solutions

1. **Cell structure and biochemical reactions**

   A

   Reaction equations:

   $$\frac{da}{dt} = k_-cd - k_+ab^2$$

   $$\frac{db}{dt} = k_-cd - k_+ab^2$$

   $$\frac{dc}{dt} = k_+ab^2 - k_-cd$$

   $$\frac{dd}{dt} = k_+ab^2 - k_-cd$$

   In steady state this gives the result in A.2.

   B

      a. Reaction equations

   $$\frac{da}{dt} = k_{-1}cd - k_{+1}ab$$

   $$\frac{dc}{dt} = k_{-2}ef - k_{+2}c$$

      b. In steady state this gives the result in B.3.

   C

   Equilibrium approximation:

   $$k_{-1}c = k_{+1}se$$

   $$K_sc = s(e_o - c)$$

   $$c = e_o\frac{s}{K_s + s}$$

   $$V = \frac{dp}{dt} = k_{+2}c = k_{+2}e_o\frac{s}{K_s + s}$$

   Quasi-steady-state approximation

   $$(k_{+2} + k_{-1})c = k_{+1}se$$

   $$K_mc = s(e_o - c)$$

   $$c = e_o\frac{s}{K_m + s}$$

   $$V = \frac{dp}{dt} = k_{+2}c = k_{+2}e_o\frac{s}{K_m + s}$$

   Since $K_m > K_s$, the reaction velocity is always smaller for the quasi-steady-state approximation.

Solutions

D

This is known as the double reciprocal form:

$$\frac{1}{V} = \frac{K}{V_{max}}\frac{1}{s} + \frac{1}{V_{max}}$$

If we plot $1/V$ against $1/s$, we get a straight line. This intercepts the x axis when $1/s = -1/K$ and the y axis when $1/V = 1/V_{max}$.

E

a. Using a logarithmic transformation, we get:

$$\ln\left(\frac{V}{V_{max} - V}\right) = n\ln s - n\ln K$$

b. If we plot $\ln\left(\frac{V}{V_{max}-V}\right)$ against $\ln s$, we get a straight line. This intercepts the x axis when $\ln s = \ln K$ and the y axis when $\ln\left(\frac{V}{V_{max}-V}\right) = n\ln K$. Plotting the values given yields values of $n = 2.394$ and $K = 1.5$ mM/s. Yes it is a good fit.

F

a. Quasi-steady-state approximation:

$$k_{+1}se = (k_{+2} + k_{-1})c_1$$
$$k_{+3}ie = k_{-3}c_2$$
$$e\left(\frac{s}{K_m} + \frac{i}{K_i} + 1\right) = e_o$$
$$V = \frac{dp}{dt} = k_{+2}c_1 = k_{+2}\frac{e_o}{\left(\frac{s}{K_m} + \frac{i}{K_i} + 1\right)}\frac{s}{K_m} = \frac{V_{max}s}{K_m(1 + i/K_i) + s}$$

b. As the inhibitor increases, the curve shifts to the right. The intercepts on the double reciprocal plot would only be affected on the x axis, where the intercept would move to the right (i.e. towards the origin).

2. **Cellular homeostasis and membrane potential**

A

a. $V = V_0$.
   The concentrations of P and Q are the same inside and outside, this is the isotonic case and the cell is already in equilibrium for concentrations and thus the volume stays the same as the original.

b. $V = 2V_0$.
   We require that $[P] = [Q]$ in equilibrium, but we have initially $[P]_0 = 2[Q]_0$, i.e. a hypertonic osmotic imbalance. Given that the volume of solution outside the cell is far larger than the cell itself (effectively infinite) the concentra-

tion of Q is not going to change, thus water must move into the cell to dilute P: to halve the concentration inside we need to double the volume. Formally, we require $[P]/V = 50$, initially we have $[P]_0/V_0 = 100$, solving for $V = 2V_0$.

B

The cell will expand indefinitely: the only way that the two requirements of equal internal and external concentration of Q and equal total concentration can be met is for the concentration of P to be zero, the cell expands to infinite size.

Formally, we have two equilibrium equations:

$[Q]_i = [Q]_e$

$[P]_i + [Q]_i = [Q]_e$

where the subscripts i and e refer to the internal and external environments respectively. The only solution is $[P]_i = 0$.

C

a. $V = V_0$
We have two equilibrium equations:
Concentration balance for R:
$[R]_i = [R]_e$

Total concentration (osmolarity):

$[P]_i + [R]_i = [Q]_e + [R]_e$

These equations are satisfied with the initial concentrations so the system is in equilibrium.

b. $V = 2 V_0$.
We have the same two equilibrium equations as in part a), but the system is not initially balanced. However
 - the total amount of P is fixed inside the cell
 - the concentrations of Q (outside the cell) is fixed
 - the concentration of R both inside and outside the cell is fixed, since the external concentration will not be altered by movement of R into the cell (at least to the same degree as the internal concentration) and R is free to move in and out of the cell.

Writing the balance of total concentration in terms of quantities of P and R and the final volume:

$P/V + R/V = 50 + 100$

Concentration of R is fixed:

$R/V = 100$

Thus:

$P/V = 50$

Since $P_0 = 100 V_0$ then $V = 2 V_0$.

D

a. Yes

b. This example is a bit different to the cell examples so far, in that we have a closed system—the external concentration is not fixed. We are also examining a system in which water isn't permeable—so we do not need to consider osmolarity and there are no other permeable uncharged species.

Using conservation of mass, since the volume is fixed this is the same as considering the total concentrations of these species:

$$[K^+]_l + [K^+]_r = 700 \text{ mM}$$
$$[Cl^-]_l + [Cl^-]_r = 300 \text{ mN}$$

Charge neutrality:

$$[K^+]_l - [Cl^-]_l - 2[X^{2-}]_l = 0$$
$$[K^+]_r - [Cl^-]_r = 0$$

Gibbs-Donnan:

$$[K^+]_r \times [Cl^-]_r = [K^+]_l \times [Cl^-]_l$$

Solving these equations gives the following equilibrium concentrations

$$[K^+]_r = 210 \text{ mM}$$
$$[K^+]_l = 490 \text{ mM}$$
$$[Cl^-]_r = 210 \text{ mM}$$
$$[Cl^-]_l = 90 \text{ mM}$$
$$[X2-]_l = 200 \text{ mM}$$

c. To achieve the correct concentration and electrical gradients for Cl^- leads to changes in the K^+ concentrations to the left and right of the membrane. As the left gets more negative and the right more positive due to Cl^- ion movement, K^+ has to migrate to the left to guarantee charge neutrality on either side.

d. $E_K = 58 \text{ mV} \times \log\left(\frac{[K^+]_r}{[K^+]_l}\right) = -21.34 \text{ mV}$

E

a. Concentration balance:

$$a + b + c + 108 = 120 + 5 + d$$

Charge balance (both inside and outside)

$$a + b = c + 108 \times \frac{11}{9}$$

$$120 + 5 = d$$

Gibbs-Donnan equilibrium:

$bc = 5 \times 125$

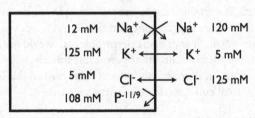

b. For potassium ions:

$$E_K = 58 \text{ mV} \times \log\left(\frac{[K^+]_o}{[K^+]_i}\right) = -81 \text{ mV}$$

For chloride ions

$$E_{Cl} = -58 \text{ mV} \times \log\left(\frac{[Cl^-]_o}{[Cl^-]_i}\right) = -81 \text{ mV}$$

As expected they are equal and the membrane potential is -81 mV.

c. For sodium ions:

$$E_{Na} = 58 \text{ mV} \times \log\left(\frac{[Na^+]_o}{[Na^+]_i}\right) = +58 \text{ mV}$$

F

a. Using the Nernst equation:

Ion	Equilibrium potential (mV)
Na^+	+67
K^+	−90
Cl^-	−34
Ca^{2+}	+118

This is not in equilibrium since they not equal. This means that the membrane potential will fall somewhere in between the extremes and be determined also by permeability.

b. Total concentration (osmotic) balance:

$$10 + 140 + 30 + 10^{-4} + [P] = 145 + 4 + 114 + 1.2$$

$$[P] = 84.2 \text{ mM}$$

Internal charge balance:

$$10 + 40 - 30 + 2 \times 10^{-4} + Z \times [P] = 0$$

$Z = -1.43$ (the net charge on the internal proteins etc).

c. External charge balance:

$$145 + 4 - 114 + 1.2 \neq 0$$

The cell cannot be in a steady state in this case as electrical neutrality outside the cell is not met.

3. **The action potential**

A

$E_m = +47.5$ mV.

B

The total current flowing from inside to outside, I_{app}, is the sum of the four branches. For sodium: the current of sodium ions arises from the difference in potential between the sodium equilibrium and the membrane potentials $(E_m - E_{Na})$ and this passes through the membrane subject to a conductivity (1/resistance) of g_{Na}, using Ohm's law:

$$i_{Na} = g_{Na}(E_m - E_{Na}).$$

Likewise for potassium and the 'leakage' current:

$$i_K = g_K(E_m - E_K),$$
$$i_L = g_L(E_m - E_L).$$

The membrane acts like a capacitor and thus current flow in response to changes in membrane potential can be described by:

$$i_m = C_m \frac{dE_m}{dt}.$$

Summing:

$$I_{app} = g_K(E_m - E_K) + g_{Na}(E_m - E_{Na}) + g_L(E_m - E_L) + C_m \frac{dE_m}{dt}.$$

Re-arranging gives:

$$C_m \frac{dE_m}{dt} = -g_K(E_m - E_K) - g_{Na}(E_m - E_{Na}) - g_L(E_m - E_L) + I_{app}.$$

C

a. The two equations are:

$$\frac{dO}{dt} = \alpha C - \beta O,$$

$$\frac{dC}{dt} = -\alpha C + \beta O.$$

b. The sum of the probabilities is 1. Hence:

$$\frac{dC}{dt} = \alpha(1 - O) - \beta O.$$

c. The steady state value and time constant are:

$$\overline{O} = \alpha/(\alpha + \beta),$$

$$\tau_O = 1/(\alpha + \beta).$$

D

a. This is just a substitution of Eqs. 3.3–3.5 into the expressions from Exercise C for each of the gates in turn.

b. In the steady-state with the potential at zero (remember that this is relative to the reference value), the h gates are largely open and the m and n gates largely closed, this requires $\beta_m \gg \alpha_m$ and $\beta_h \ll \alpha_h$.

   When the voltage increases, the gates go from open to closed and vice versa. The time constants for the n and h gates are much larger than for the m gate, so the m gate responds much more rapidly, as expected, this requires $\beta_m \gg \alpha_m$ and $\beta_h \ll \alpha_h$.

E

About 2.3 μA.

F

a. If the permeability to K^+ is high compared to other ions then E_{K^+} will dominate the membrane potential. From exercise 2F we know that E_{K^+} is -90 mV so the range suggested seems reasonable.

b. From the figure the membrane potential plateaus at $+52$ mV. If, as suggested, we ignore the other ions and assume that Ca^{2+} permeability dominates then we need $E_{Ca^{2+}}$ equal to or greater than $+52$ mV:

$$E_{Ca^{2+}} = \frac{58 \text{ mV}}{2} \log_{10}\left(\frac{1.2}{[Ca^{2+}]_i}\right) \geq +52 \text{ mV}.$$

This requires that $[Ca^{2+}]_i \leq 0.019$ mM.

4. Cellular transport and communication

A

Starting with the diffusion equation in 1-dimension:

$$\frac{\partial c}{\partial t} = f + D\frac{\partial^2 c}{\partial x^2}.$$

We are looking for transport in the steady state, thus changes with respect to time are zero. We will also assume there is no generation (or consumption) of the species within the membrane thus $f = 0$:

$$D\frac{\partial^2 c}{\partial x^2} = 0.$$

This is a standard second order differential equation with a solution of the form $c = ax + b$. The boundary conditions from the diagram are:

$$c(x = 0) = c_o,$$
$$c(x = L) = c_i.$$

Hence

$$c = c_o + (c_i - c_o)\frac{x}{L}.$$

Using Fick's law we can write the flux:

$$J = D\frac{\partial c}{\partial x} = \frac{D}{L}(c_i - c_o).$$

If we were to account for relative solubility via the partition coefficient then the boundary conditions would change—since the concentration at the boundaries **inside** the membrane would be higher (or lower) compared to the same boundary outside the membrane:

$$c(x = 0) = Kc_o.$$
$$c(x = L) = Kc_i.$$

The solution would look like:

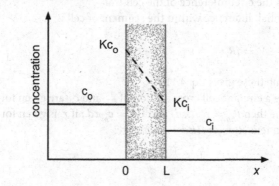

B

The differential equations for this process are:

$$\frac{dp_i}{dt} = kp_e - kp_i + k_+ s_i c_i - k_- p_i,$$

$$\frac{dp_e}{dt} = kp_i - kp_e + k_+ s_e c_e - k_- p_e,$$

$$\frac{dc_i}{dt} = kc_e - kc_i + k_- p_i + k_+ s_i c_i,$$

$$\frac{dc_e}{dt} = kc_i - kc_e + k_- p_e - k_+ s_e c_e.$$

C

a. This constant current source can equally well be simulated in the equations by an increase in the resting potential for potassium, since both contribute a fixed term on the right hand side of the equation. A change in equilibrium potential is directly related to concentration via the Nernst equation. The necessary rise in potassium equilibrium potential is:

$$v_K = \frac{I_{app}}{\bar{g}_K n^4} = \frac{2.3}{36 \times 0.3177^4} = 6.27 \text{ mV}.$$

b. Relating the potassium concentration to potential by the Nernst equation, the relative change in extracellular potassium is given by:

$$\left([K^+]_o\right)_{new} = \left([K^+]_o\right)_{old} 10^{(6.27/58)} = 1.28 \left([K^+]_o\right)_{old}.$$

A similar approach for sodium gives a factor of over 500, which is not physiologically possible!

D

a. The current flowing in the membrane is given by:

$$i + \frac{\partial i}{\partial x}\delta x - i = C_m c \delta x \frac{\partial V}{\partial t} + \frac{c \delta x}{R_m} V,$$

where c is the circumference of the cell wall.
The potential difference within the segment of cell is:

$$V + \frac{\partial V}{\partial x} - V = iR_c.$$

Combining these gives Eq. 4.11.

b. Assuming a circular cell cross section: if r_m and c_m are given for a unit area of membrane then $R_m = r_m/(\pi d)$ and $C_m = c_m \pi d$; if r_c is given for a unit area of cytoplasm tens $R_c = 4r_c/(\pi d^2)$. Therefore:

$$\tau_m = r_m c_m.$$

$$\lambda_m = \sqrt{\frac{r_m d}{4 r_c}}.$$

c. $\tau_m = 8.4\,\text{ms}$ and $\lambda_m = 0.15\,\text{cm}$.

d. This is a second order differential equation in x with solution (including boundary condition at $x = 0$) of: $V = 100\,e^{-x/\lambda_m}$. Solving for x when $V = 20\,\text{mV}$ gives $x = 2.4\,\text{mm}$.

5. Pharmacokinetics

A

a. Starting with:

$$V_c \frac{dC_p}{dt} = -V_c k_e C_p(t) + A_{in}(t).$$

Take the Laplace transform:

$$s V_c C_p(s) = -V_c k_e C_p(s) + A_{in}(s).$$

Rearrange

$$C_p(s) = \frac{1}{V_c} \frac{A_{in}(s)}{(s + k_e)}.$$

Inverse Laplace transform (using the convolution property)

$$C_p(t) = \frac{1}{V_c} \int_0^t A_{in}(\lambda) e^{-k_e(\lambda - t)} d\lambda.$$

b. $A_{in}(t) = k_0$

$$C_p(t) = \frac{k_0 e^{-k_e t}}{V_c} \int_0^t e^{k_e \lambda} d\lambda,$$

$$C_p(t) = \frac{k_0 e^{-k_e t}}{V_c k_e} \left(e^{k_e t} - 1 \right),$$

$$C_p(t) = \frac{k_0}{V_c k_e} \left(1 - e^{-k_e t} \right).$$

c. The steady state concentration can be found in the limit $t \to \infty$

$$C_p^\infty = \frac{k_0}{V_c k_e}.$$

As might be expected this concentration depends on the relative rate of elimination and delivery, as well as the volume of distribution into which the substance is dissolved.

B

Starting with:

$$\frac{dA_a}{dt} = -k_a A_a(t) + A_{in}(t),$$

$$\frac{dA_p}{dt} = k_a A_a(t) - k_e A_p(t).$$

a. Laplace transform (with zero initial conditions):

$$A_a(s) = \frac{A_{in}(s)}{s + k_a},$$

$$A_p(s) = \frac{k_a A_a(s)}{s + k_e}.$$

Thus:

$$A_p(s) = \frac{k_a A_{in}(s)}{(s + k_e)(s + k_a)}.$$

Inverse Laplace transform:

$$A_p(t) = \frac{k_a}{k_a - k_e}(e^{-k_e t} - e^{-k_a t}) * A_{in}(t).$$

$A_{in}(t) = D * \delta(t)$, where D is the dose.

Also $A_p = C_p * V_c$:

$$C_p(t) = \frac{D}{V_c} \frac{k_a}{k_a - k_e}(e^{-k_e t} - e^{-k_a t}).$$

b. A sum of a growth toward saturation plus a decay toward zero leads to something that firstly increases and then decays away.

c. The slowest time constant, which we would expect to be elimination, will dominate at later time points and thus a straight line fit to the log values will give an estimate for this time const.

C

a. Using the data in the table we need to calculate the area under the curve for both methods of administration. Trapezoidal integration will suffice for this calculation. We do not have measurements at very long time post-administration so we do not know when the concentration drops to zero, we will ignore any concentration after the last time point:

$$F = \frac{AUC_{oral}}{AUC_{IV}} = \frac{251.07}{386.04} = 0.650$$

b. We can use the IV injection to quantify the elimination kinetics, it will obey Eq. 5.6. We can estimate the parameters either from a semi-logarithmic plot of concentration versus time, which gives a straight line slope of -0.500 and intercept of 5.298, or we could use non-linear model fitting of the model to the data:

Solutions

$k_e = 0.500 \text{ h}^{-1}.$

$T_{1/2} = \dfrac{\ln 2}{k_e} = 1.386 \text{ h}.$

For the absorption constant, and thus the half-life, we need to use Eq. 5.11, where we have already obtained k_e. Again we could use model fitting of the equation to the data, or we could employ the 'method of residuals'. Writing Eq. 5.11 as:

$C_p = Ae^{-k_e t} - Ae^{-k_a t}.$

If $k_a > k_e$, then as $t \to \infty$

$C_p \approx Ae^{-k_e t}.$

We apply this to the final three time points of data: a semi-log plot allows us to calculate A and k_e from a slope of -0.500 and intercept of 5.00. Note that we get the same value for k_e as above (unsurprisingly) but the intercept value is different because of the bioavailability.

Now define the 'residual':

$R = Ae^{-k_e t} - C_p.$

We have all the information we need to calculate R at every time point and by definition:

$R = Ae^{-k_a t}.$

Thus a semi-log plot of R against time will allow k_a to be estimated from the slope. Using the first three data points (where the residual will be the largest, since absorption dominates here) gives a slope of -3.03. Hence:

$k_a = 3.03 \text{ h}^{-1},$

$T_{1/2} = \dfrac{\ln 2}{k_a} = 0.229 \text{ h}.$

We should check at this point that $k_a > k_e$, which it is and sufficiently so that the assumption we made above is reasonable.

c. To calculate the volume of distribution we need the initial plasma concentration. We can use the data we have in part (a), for example using the intercept from the IV administration.

$C_p^0 = 199.94 \text{ mg/L}.$

Thus

$V_c = \dfrac{D}{C_p^0} = 10.0 \text{ L}.$

Which is larger than the total plasma volume, so must include some fast exchanging tissue.

D

a. We need Eq. 5.11

$$C_p(t) = \frac{FD}{V_c} \frac{k_a}{k_a - k_e} (e^{-k_e t} - e^{-k_a t}).$$

b. We can find the maximum dose by firstly finding the maximum concentration, for which we will need to determine the time of maximum concentration by finding the turning point of the function described by C_p. This gives

$$t_{max} = \frac{1}{k_a - k_e} \ln \left(\frac{k_a}{k_e} \right) = 21.7 \text{ min.}$$

Substituting into the expression for C_p and then converting to dose

$$D_{max} = \frac{V_c C_p^{max}}{F} = 1330 \text{ mg.}$$

It is possible to find a nice simplified form for the maximum concentration with a bit of effort:

$$C_p^{max} = \frac{FD}{V_c} \left(\frac{k_a}{k_e} \right)^{-k_e/k_a - k_e}.$$

c. The drug is no longer effective once $C_p < X$ where $X = 20$ mg/L, thus we need to solve for the time at which:

$$C_p(t) = \frac{FD}{V_c} \frac{k_a}{k_a - k_e} (e^{-k_e t} - e^{-k_a t}) = X.$$

Noting that $k_e \ll k_a$, the absorption process is very rapid in comparison to elimination, so we might assume there is no further absorption by the time that the concentration drops below the level for efficacy. We can thus simplify the expression assuming that the exponential term containing k_a is zero:

$$\frac{FD}{V_c} \frac{k_a}{k_a - k_e} e^{-k_e t} = X.$$

Giving:

$$t = -\frac{1}{k_e} \ln \left(\frac{k_a - k_e}{k_a} \frac{V_c}{FD} X \right) = 606 \text{ min.}$$

We could check our assumption from above that absorption isn't important by this time by comparing 606 min to the time constant for absorption ($1/k_a$) which is approximately 5 min.

d. Michaelis-Menten kinetics are of the form:

$$\frac{V_{max} C}{C + K_m}.$$

This is linear if $K_m \gg C$, first order kinetics, but is constant if $K_m \ll C$, zeroth order kinetics. In this case K_m is an order of magnitude greater than C at any point in time, given that $C_{p,max}$ is around 100 mg/L. So we would conclude that the analysis still holds as the elimination would still look like first order and has not saturated.

6. Tissue mechanics

A

a. The one-dimensional equations are:

$$\frac{\partial \sigma_x}{\partial x} - \rho g = 0$$

$$\varepsilon_x = \frac{\partial u}{\partial x}$$

$$\varepsilon_x = \frac{\sigma_x}{E}$$

Hence:

$$\frac{\partial^2 u}{\partial x^2} = \frac{\rho g}{E}$$

Integrate twice, and substitute in the boundary conditions given to derive the answer.

b. At the top of the tissue:

$$u = \frac{\rho g L^2}{2E}$$

This is approximately 1 mm (10% of the tissue height) for the values given.

B

a. Re-arrange and square the result:

$$\left(\frac{3K\varepsilon}{RT} + c^* \right)^2 = c^{*2} + \left(\frac{\phi_0^w c_0^f}{3\varepsilon + \phi_0^w} \right)^2$$

Then multiply out the LHS and re-arrange to get the answer given.

b. The result is obviously very non-linear, with concentration becoming very high at low values of pressure.

C

a. The one-dimensional equations are:

$$G \frac{\partial^2 u}{\partial x^2} + \frac{G}{1 - 2v} \frac{\partial \varepsilon}{\partial x} = \alpha \frac{\partial p}{\partial x}$$

$$\varepsilon = \frac{\partial u}{\partial x}$$

$$\frac{\kappa}{\mu} \frac{\partial^2 p}{\partial x^2} = \alpha \frac{\partial \varepsilon}{\partial t} + \frac{1}{Q} \frac{\partial p}{\partial t}$$

Integrate the first equation wrt x:

$$\frac{2G(1-v)}{\alpha(1-2v)}\varepsilon = p$$

Substitute into the last equation:

$$\frac{\kappa}{\mu}\frac{\partial^2 p}{\partial x^2} = \left(\frac{\alpha^2(1-2v)}{2G(1-v)} + \frac{1}{Q}\right)\frac{\partial p}{\partial t}$$

Hence the expression given, where the coefficient is:

$$\frac{1}{c} = \frac{\mu}{\kappa}\left(\frac{\alpha^2(1-2v)}{2G(1-v)} + \frac{1}{Q}\right)$$

b. Easiest way is to take Laplace transforms (with zero initial condition):

$$\frac{\partial^2 p}{\partial x^2} - \frac{s}{c}p = 0$$

Solve, ignoring the positive exponential term (which would tend to infinity):

$$p = Ae^{-\sqrt{\frac{s}{c}}x}$$

Use the boundary condition (Laplace transformed):

$$p = \frac{P}{s}e^{-\sqrt{\frac{s}{c}}x}$$

This can be inverse Laplace transformed (using a standard result) to give:

$$p = P\left(1 - \text{erf}\left(\frac{x}{2\sqrt{ct}}\right)\right)$$

D

This can be solved by integrating up in stages to give:

$$\frac{d}{dr}\left(r\frac{dc}{dr}\right) = \frac{Mr}{D}$$

$$r\frac{dc}{dr} = \frac{Mr^2}{2D} + A$$

$$c = \frac{Mr^2}{4D} + A\ln r + B$$

Then insert the two boundary conditions to give:

$$c = c_o + \frac{M}{4D}\left(r^2 - R_i^2\right) + \frac{MR_i^2}{2D}\ln\left(\frac{R_i}{r}\right)$$

7. Cardiovascular system I

A

For the numbers given, cardiac output would be 200/(0.21−0.16) = 4000 ml_blood/minute. Stroke volume = 4000/60 = 67 ml_blood.

B

    a. The RR interval is 0.8 s, so the heart rate = 60/0.8 = 75 bpm. There is no variability shown here, which is not realistic.

    b. You should find that it goes up quite rapidly, the additional cardiac output providing more oxygen to your muscles.

8. Cardiovascular system II

A

Integrate up the equation wrt radius twice:

$$\mu u = \frac{r^2}{4} \frac{dp}{dx} + A \ln r + B$$

For the solution to be finite at the origin, $A = 0$. Also the velocity is zero at the wall, hence:

$$u = \frac{1}{4\mu}(r^2 - R^2)\frac{dp}{dx}$$

Flow rate:

$$q = \int_{r=0}^{R} u(r)dA = \int_{r=0}^{R} \frac{1}{4\mu}(r^2 - R^2)\frac{dp}{dx}2\pi r dr = \frac{dp}{dx}\frac{\pi R^4}{8\mu}$$

Since the pressure gradient is the pressure drop divided by the vessel length, we get:

$$\frac{\Delta p}{q} = \frac{8\mu L}{\pi R^4}$$

B

When placed in parallel, the pressure difference is the same for each vessel, with the total flow rate being the sum of the individual flow rates. Hence:

$$q = \sum_{i=1}^{N} \frac{\Delta p}{\mathcal{R}} = \Delta p \sum_{i=1}^{N} \frac{1}{\mathcal{R}} = \Delta p \frac{N}{\mathcal{R}}$$

The overall resistance is thus $\mathcal{R}/N$.

C

a. It is easiest to work in terms of infinitesimal strains, since compliance is defined as the rate of change of volume with pressure:

$$C = \frac{dV}{dp} = L\frac{dA}{dp} = 2\pi RL\frac{dR}{dp} = \frac{2\pi R^2 L}{dp}\frac{dR}{R} = \frac{2\pi R^2 L}{dp}d\varepsilon_\theta$$

Substitute in Hooke's law in polar co-ordinates:

$$C = \frac{2\pi R^2 L}{dp}\frac{1}{E}(d\sigma_\theta - vd\sigma_r - vd\sigma_z)$$

Substitute in the stress components to give:

$$C = \frac{2\pi R^3 L}{Eh}\left(1 - \frac{v}{2}\right)$$

This gives the quoted result, since Poisson's ratio is equal to 1/2.

b. When placed in parallel, capacitors essentially simply become one large capacitor with storage capacity (capacitance) adding. Hence N capacitors have capacitance CN. The same applies for compliance.

D

You should be able to calculate the values easily.

E

a. Denoting the pressure at entrance to the capillary compartment by P_a, conservation of current/flow at this node gives:

$$\frac{P_{in} - P_a}{R_a + i\omega I_a} + \frac{0 - P_a}{R_c + R_v + i\omega I_v} + \frac{0 - P_a}{1/i\omega C_a} = 0$$

The flow through the capillaries is then:

$$q = \frac{P_a}{R_c + R_v + i\omega I_v}$$

$$= \frac{P_{in}}{(R_a + i\omega I_a + R_c + R_v + i\omega I_v + i\omega C_a(R_a + i\omega I_a)(R_c + R_v + i\omega I_v))}$$

b. This is of the form of a (third-order) low-pass filter. This is important because it filters out the pulsatile nature of the flow in the aorta such that flow is steady by the time that it reaches the capillaries.

F

Boundary conditions are:

$$\sigma_r|_{r=R} = -p$$
$$\sigma_r|_{r=R+h} = -p_{ext}$$

These are obviously satisfied.

G

a. The mean flow velocity is found from:

$$\pi R^2 U = \int_{r=0}^{R} u(r) dA$$

$$U = \frac{2U_{max}}{R^2} \int_{r=0}^{R} \left(1 - \left(\frac{r}{R}\right)^n\right) r dr$$

$$U = U_{max}\left(\frac{n}{n+2}\right)$$

b. The frictional force is given by:

$$f = 2\pi R \mu \left.\frac{\partial u}{\partial r}\right|_{r=R} = -2\pi n\mu U_{max} = -2\pi(n+2)\mu U$$

H

Eq. 8.29 shows that inductance can be neglected even in the larger vessels that we are considering here. The transfer function thus simplifies to:

$$q = \frac{P_{in}}{(R_a + R_c + R_v + i\omega C_a R_a(R_c + R_v))}$$

This has time constant:

$$\tau = \frac{C_a R_a(R_c + R_v)}{R_a + R_c + R_v}$$

Any frequencies above the cut-off frequency set by this time constant will be filtered out, hence we need:

$$2\pi > \frac{R_a + R_c + R_v}{C_a R_a(R_c + R_v)}$$

I

a. We know that flow is proportional to the product of pressure difference (which will be P here) and radius to the power four:

$$q = k_1 P R^4$$

If compliance is constant, then the volume is proportional to pressure:

$$V = k_2 P$$

Since volume is proportional to radius squared:

$$P = k_3 R^2$$

Putting these all together gives:

$$q = k_1 k_3 R^6$$

b. Force balance on the fluid gives wall shear stress as:

$$\tau_w 2\pi R = \frac{dp}{dx}\pi R^2$$

Hence:

$$\tau_w = \frac{dp}{dx}\frac{R}{2}$$

Hence:

$$\tau_w = k_4 R^3$$

J

a. If the vessel is incompressible, then the cross-sectional area must be the same at all times:

$$(R + h)^2 - R^2 = (R_o + h_o)^2 - R_o^2$$

Hence:

$$h = -R + \sqrt{R^2 + 2R_o h_o + h_o^2}$$

b. Clearly, as R increases, h decreases. Resistance is simply inversely proportional to vessel radius to the power four, so it drops very rapidly with increased radius.
c. Compliance is proportional to radius to the power three and inversely proportional to wall thickness, so as radius increases, compliance also goes up very rapidly.

9. The respiratory system

A

In the steady state with no generation or consumption of the gas involved the diffusion equation for the partial pressure of gas in the membrane is:

$$D\frac{\partial^2 c}{\partial x^2} = 0.$$

The boundary conditions are:

- Concentration of the gas on the airspace side of the membrane ($x = 0$). For this we need to use the Ostwald co-efficient to 'convert' partial pressure of gas into a concentration: $c(x = 0) = \sigma p_g$.
- Concentration of gas in the blood: $c(x = L) = c_b$.
  Solving and using Fick's law the flux is thus:

$$q = \frac{D}{L}(c_b - \sigma p_g).$$

We can define D/L as the surface diffusion co-efficient D_s, which is a property of the membrane.

B

a. $-f\frac{dC(x)}{dx} = 2\pi rD_s(C(x) - C_o)$

b. The equation is a first order differential equation, so can be written in the form:

$$\frac{F}{2\pi rD_s}\frac{dC}{dx} + C = C_o$$

which has the solution (given the boundary condition):

$$C = C_o + (C_{in} - C_o)\exp\left(-\frac{2\pi rD_s}{F}\right)$$

c. The flux is found by integrating the function above

$$Q = \int\limits_{x=0}^{L} 2\pi rD_s(C(x) - C_o)dx$$

to give:

$$Q = (C_{in} - C_o)f\left[1 - \exp\left(-\frac{2\pi rD_sL}{F}\right)\right]$$

d. The flux is 4×10^{-14} mM/s for baseline flow, 3.2×10^{-14} mM/s and 4.5×10^{-14} mM/s for halved and doubled flow. Hence it is relatively constant with flow rate in this region (flux saturates at high flows, increasing proportionally less than flow).

e. Two competing effects: increasing the flow increases the oxygen flux into the capillary, but it passes through the vessel more quickly, so has less chance to diffuse into the surrounding tissue.

f. This gives Eq. 9.5, the equation no longer depends upon the diffusion properties of the capillary wall only on perfusion and concentration (partial pressure) differences. A large surface area for gas transfer is present in the lungs and the membrane is thin, thus diffusion doesn't limit the overall transfer of gas, thus L is large compared to the diffusion 'distances' involved.

C

a. Using reaction kinetic principles from ▶ Chap. 1, the equilibrium constant for the second part of the reaction scheme is given by:

$$K_A = \frac{[HCO_3^-][H^+]}{[H_2CO_3]}.$$

b. If almost all the available CO_2 converts to carbonic acid, then the concentration of CO_2 is very similar to that of H_2CO_3, thus the 'corrected' equilibrium constant is:

$$K_A = \frac{[HCO_3^-][H^+]}{[CO_2]}.$$

D

a. In Exercise 9B the governing equation for gas concentration in the capillary blood was:

$$-F\frac{dC(x)}{dx} = 2\pi rD_s(C(x) - C_o)$$

This can be applied to CO_2, but we now need to account for the conversion to bicarbonate. As suggested, we can model this using a simplified reaction scheme that ignores the intermediary H_2CO_3:

$$CO_2 + H_2O \underset{k_{-1}}{\overset{k_1}{\rightleftharpoons}} HCO_3^- + H^+$$

Including the 'loss' of CO_2 to bicarbonate and the reverse process where bicarbonate converts back to CO_2:

$$F\frac{dC}{dx} = 2\pi rD_s(\sigma_{CO_2}P_{CO_2} - C) + k_{-1}[H^+]D - k_1C,$$

where we use D to signify $[HCO_3-]$ as we have used C to represent $[CO_2]$. We also have an equation for D, but as this exists only in solution we do not have to worry about the gas exchange part only the reaction kinetics:

$$F\frac{dD}{dx} = k_1C - k_{-1}[H^+]D.$$

b. Combining the two equations gives:

$$F\frac{d}{dx}(C + D) = 2\pi rD_s(\sigma_{CO_2}P_{CO_2} - C).$$

c. Taking a quasi-steady state approximation allows us to write $D = K_aC$ where $K_c = k_1/(k_{-1}[H^+])$. Thus, if $[H^+]$ is constant with respect to x:

$$F(1 + K_c)\frac{dC}{dx} = 2\pi rD_s(\sigma_{CO_2}P_{CO_2} - C).$$

This is practically identical to the result in Exercise 9B and thus the solution for flux in the limit will be the same except for an 'amplification' of $(1+K_c)$, hence Eq. 9.7.

d. We can find the relationship between K_c and K_a using the result in Exercise 9C:

$$K_a = K_c[H^+]$$

Since $\log_{10}K_a = -6.1$ and $pH = -\log_{10}[H^+] = 7.4$ typically, K_c is around 20. A substantial amplification of the flux of CO_2 allowing more rapid removal from the body than would be achievable by having CO_2 alone in solution.

E

a. This is a matter of taking logs of the expression from Exercise 9C part b) and re-arranging in terms of $-\log[H^+]$, i.e. pH.

b. For the values given, $[CO_2] = 1.2$ mM and pH $= 7.4$.

c. This is called the Davenport diagram and is shown below. Large changes in pH can be gained by adjusting both pCO_2 and bicarbonate concentration, so the lungs and kidneys have a lot of control over pH.

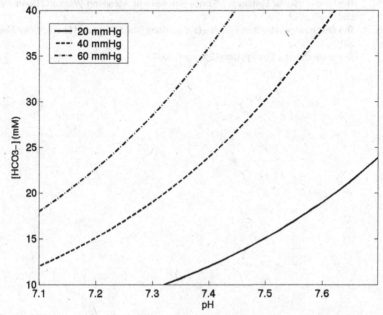

d. The kidneys need to excrete H^+ and reabsorb HCO_3^- to balance pH again. This would happen in a person with poor respiration (not enough removal of CO_2 from the lungs).

Further Reading

1. Cardiovascular Physiology (8th edition): Mohrman, Heller. McGraw-Hill, 2013.
2. Cellular Physiology of Nerve and Muscle (4th edition): Matthews. Blackwell, 2002.
3. Mathematical Physiology (2nd edition): Keener, Sneyd. Springer, 2008.
4. Molecular Biology of the Cell (6th edition): Alberts, Johnson, Lewis, Morgan, Raff, Roberts and Walter. Garland Science, 2014.
5. The Cardiovascular System at a Glance (4th edition): Aaronson, Ward and Connolly. Wiley-Blackwell, 2012.
6. The Respiratory System at a Glance (3rd edition): Ward, Ward and Leach. Wiley-Blackwell, 2010.
7. Tissue Mechanics: Cotwin, Doty. Springer, 2007.

Printed in the United States
by Baker & Taylor Publisher Services

Printed in the United States
by Baker & Taylor Publisher Services